BOSTON STUDIES IN THE PHILOSOPHY OF SCIENCE

VOLUME X

SCIENTIFIC PROCEDURES

SYNTHESE LIBRARY

MONOGRAPHS ON EPISTEMOLOGY,

LOGIC, METHODOLOGY, PHILOSOPHY OF SCIENCE,

SOCIOLOGY OF SCIENCE AND OF KNOWLEDGE,

AND ON THE MATHEMATICAL METHODS OF

SOCIAL AND BEHAVIORAL SCIENCES

BOSTON STUDIES IN THE PHILOSOPHY OF SCIENCE

VOLUME X

EDITED BY ROBERT S. COHEN AND MARX W. WARTOFSKY

LADISLAV TONDL

SCIENTIFIC PROCEDURES

A CONTRIBUTION CONCERNING
THE METHODOLOGICAL PROBLEMS OF SCIENTIFIC CONCEPTS
AND SCIENTIFIC EXPLANATION

Translated from the Czech by David Short

D. REIDEL PUBLISHING COMPANY

DORDRECHT-HOLLAND/BOSTON-U.S.A.

Library of Congress Catalog Card Number 72–77880

ISBN 90 277 0147 4

Published by D. Reidel Publishing Company,
P.O. Box 17, Dordrecht, Holland

Sold and distributed in the U.S.A., Canada, and Mexico
by D. Reidel Publishing Company, Inc.
306 Dartmouth Street, Boston,
Mass. 02116, U.S.A.

Printed in The Netherlands by D. Reidel, Dordrecht

EDITORIAL INTRODUCTION

For a decade, we have admired the incisive and broadly informed works of Ladislav Tondl on the foundations of science. Now it is indeed a pleasure to include this book among the *Boston Studies in the Philosophy of Science*. We hope that it will help to deepen the collaborative scholarship of scientists and philosophers in Czechoslovakia with the English-reading scholars of the world.

Professor Ladislav Tondl was born in 1924, and completed his higher education at the Charles University in Prague. His doctorate was granted by the Institute of Information Theory and Automation. He was a professor and scientific research worker at the Institute for the Theory and Methodology of Science, which was a component part of the Czechoslovak Academy of Sciences. Tondl's principal fields of interest are the methodology of the empirical and experimental sciences, logical semantics, and cybernetics.

For many years, he collaborated with Professor Albert Perez and others at the Institute of Information Theory and Automation in Prague, and he has undertaken fruitful collaboration with logicians in the Soviet and Polish schools, and been influenced by the Finnish logicians as well, among them Jaakko Hintikka.

We list below a selection of his main publications. Perhaps the most accessible in presenting his central conception of the relationship between modern information theory and the methodology of the sciences is his 1965 paper with Perez, 'On the Role of Information Theory in Certain Scientific Procedures'.

Tondl has also undertaken studies in the field which has been called 'the science of science', with conscientious attention to the humane functions which may be carried out by scientific planning and by the planning of science. He has also taken a leading role in educational and publishing activities within his fields of interest. Most important perhaps is the quarterly journal *Teorie a Metoda* (Theory and Method) published by Tondl's former Institute for Theory and Methodology of Science, which began with its first volume in 1969 and has continued under the able editorship of Karel Berka.

Tondl is a member of the editorial boards of several international journals, *Synthese*, *Theory and Decision*, and *Problems of the Science of Science*.

Boston University Center for the Philosophy and History of Science
Summer 1972

R. S. COHEN
M. W. WARTOFSKY

Selected works by L. Tondl:

1957: 'Gnoseologická úloha abstrakce' ('The Epistemological Problem of Abstraction'), Sbor: *Otázky teorie poznání* (Essays [by various authors]: *Problems of the Theory of Knowledge*), Prague.

1958: 'Novopositivismus', in *Současná Západní Filosofie (Contemporary Western Philosophy)* (a volume in the *Malá Moderní Encyklopedie*), Orbis, Prague, pp. 17–54.

1959: 'Kauzální analýza a kauzální explikace', ('Causal Analysis and Causal Explanation'), in *K metodologii experimentálních věd, (Toward the Methodology of Experimental Science)*, (with Otakar Zich and Ivan Málek), Czechoslovak Academy of Sciences, Prague.

1960: 'The Cognitive Role of Abstraction' in *Methodological and Philosophical Problems of Scientific Abstraction*, Moscow (in Russian).

1965: (a) 'Aspects du problème de reduction en science du point de vue de la théorie de l'information' (with Albert Perez), *Le concept de l'information dans la science contemporaine*, Gauthier-Villars, Paris.

(b) 'On the Role of Information Theory in Certain Scientific Procedures' (with Albert Perez), *Information and Prediction in Science*, ed. Dockx and Bernays, Academic Press, New York.

(c) 'Modely některých vědeckých procedur z hlediska logiky a teorie informace' ('Models of Some Scientific Procedures from the Standpoint of Logic and Information Theory') Sbor: *Problémy kybernetiky*, Nakl. ČSAV, Prague.

1966: (a) *Problémy Sémantiky, (Problems of Semantics)*, Academia, Czechoslovak Academy of Sciences, Prague, 366 pp. (with brief resumé in English).

(b) 'Zum Identifizierungsproblem in den empirischen Wissenschaften', *Rostocker Philosophische Manuskripte*, No. 3 (University of Rostock), pp. 51–55.

(c) 'Concerning the Notions of "Technics" and "Technical Science"', *Organon* **3**, pp. 111–125 (in Russian).

(d) *Vybrané kapitoly o společenské funkci vědy (Selected Papers on the Social Function of Science)*, Czechoslovak Academy of Sciences, Prague, 256 pp. (the thirteen chapters of this book gather together Tondl's principal papers from 1962–66 on the social functions of science form a variety of philosophical, technical, and general publications).

1969: (a) 'On the Position and Task of Science in the Scientific and Technological Revolution', *Yearbook 67 of the Czechoslovak Academy of Sciences*, Prague, pp. 7–34 (in English).
(b) *Man and Science*, Institute for the Theory and Methodology of Science of the Czechoslovak Academy of Sciences, Prague, 128 pp. (in English).
(c) 'Socio-Economic Background of Scientific Activity', *Transfer of Research Results in Social-Economic Practice* (Collection of Papers), Institute for the Theory and Methodology of Science, Czechoslovak Academy of Science, Prague, pp. 4–37 (in English; this volume comprised the papers of a Czechoslovak-Swedish symposium of 1969).
(d) 'Scientific Explanation' *TAM (Teorie a Metoda)* **1**, No. 1, pp. 3–43 (in English).
(e) 'Socio-Economic Background of Scientific Activity', *TAM* **1**, No. 2, pp. 9–42 (in English).
(f) 'Empirické východisko a analýza "universe de discours"' ('The Empirical Starting Point and the Analysis of the "universe of discourse"'), *TAM* **1**, No. 3, pp. 11–35 (in Czech).
(g) 'Society and Scientific Communities', *Nová Mysl*, No. 4 (in Czech).

1970: (a) 'Science as a Vocation', *TAM* **2**, No. 1, pp. 15–33 (in English).
(b) 'Předpoklady kvantifikace v empirických vědách' ('Presuppositions of Quantification in Empirical Sciences'), *TAM* **2**, No. 2, pp. 52–17 (in Czech).
(c) *Zur Konstruktion eines Systems für eine logische Darstellung der Identifizierung, Systemanalyse und Informationsverarbeitung*, Oldenburg Verlag, Munich.

1971: (a) 'A Decision Model of Scientific Law', *TAM* **3**, No. 1 pp. 71–82 (in English).
(b) 'Science, Applications, and Solutions', *TAM* **3**, No. 2, pp. 89–100 (in Russian).
(c) 'Leibnizova metoda identifikace nerozlišitelného a metody redukce rozlišovacích kritérií' ('Leibniz's Method of Identification of Indiscernibles and the Method of Reduction of Differentiation Criteria'), *TAM* **3**, No. 3, pp. 35–42 (in Czech).
(d) 'Logické schéma problémové situace' ('Logical Scheme for Problem-Solving Situations'), *Československá Psychologie* **15**, No. 6, pp. 549–560 (in Czech).

1972: (a) 'The Problems of Semantic Information', *Kybernetika* **8**, No. 3, pp. 189–212 (in Russian).
(b) 'Prerequisites for Quantification in the Empirical Sciences' (in symposium on Problems of Scientific Languages), *Theory and Decision* **2**, pp. 238–261.

PREFACE

This work sets out from the viewpoint which might be described as an informational or communicational conception of scientific activity. In this viewpoint, scientific activity is conceived as a negentropic activity *sui generis*, i.e. an activity leading towards a diminishment of the uncertainty rate in decision-making, not only in the sphere of science alone but also in that of the multifarious applications of achievements made in and by science. Scientific procedures are then characterized as operations with data, the solution of problems or the answering of questions. The actual formulation of this viewpoint has been greatly influenced by the many years I have spent working side by side with my colleagues at the Institute of Information and Automation Theory of the Czechoslovak Academy of Sciences. At the same time this viewpoint is quite close to the methods being developed by the present Finnish school in logic and the methodology of science.

A conception which accentuates the procedural approach and which relies on certain of the principles of information theory naturally differs considerably from the static conception of logico-syntactic analysis as developed by the analytical and positivist schools. For this reason, though only to the degree deemed the essential minimum, this work also contains some critical comment on the results attained by these schools. However, since an exhaustive critique of these trends is not the actual aim of the present work, which aims rather at offering a positive solution of its own, these critical passages are confined to those conceptions which are connected directly with the questions analyzed here.

The core of the work is an analysis of the problems of scientific explanation and certain related procedures. This also means that it is not and cannot be an analysis of the whole range of empirical procedures, but that it is restricted only to those of scientific explanation. The chapters on the empirical starting-point and the concepts of the language of science are also adapted to the analysis of the procedures of scientific explanation in the empirical sciences. It might then be objected that a more fitting title to the book would have been simply 'Scientific Explanation'. However, I am convinced that many of the methods I have used when analyzing the procedures of explanation may also be profitable in the analysis of other scientific procedures.

The work makes use of those methods and means of expression which are current in the contemporary methodology of the empirical sciences, mathematical logic and logical semantics. Therefore, and this concerns first and foremost the formalized sections, the reader is assumed to possess at least a minimal knowledge of these means. On the other hand the author has not selected the strict formal approach (that is in the external presentation, the way the individual formulae are given, etc., although in places this would have been possible and doubtless quite appropriate) but has proceeded in a way which endeavours not to lose sight of the potentials of intuitive comprehension and interpretation.

LADISLAV TONDL

TABLE OF CONTENTS

CHAPTER I

THE CONCEPT OF 'SCIENCE' AS COGNITIVE ACTIVITY

1. Aspects of the Concept of Science

The concept of science is a typically ambiguous concept. Likewise the attribute 'scientific' can be applied to quite heterogeneous, albeit more or less connected phenomena. For these reasons it seems expedient to clarify certain fundamental aspects of the concept of 'science'.

(a) By science we understand primarily conscious and organized *cognitive activity*. The concept of 'cognition' is of course considerably broader than that of 'science', therefore it is necessary to discriminate between scientific cognition and cognition generally. Scientific cognition is associated with a set of certain specified aims, which we shall be endeavouring to outline more clearly in the following section. The most significant aims of science are connected with what has been characterized as the *cognitive* or epistemic function of science, whereby science concentrates closest attention on the cognition of new and previously unknown scientific laws, or on the refinement of the current state of knowledge of such laws.

The cognitive function of science cannot be understood merely as the cognition of the actual world around us. The center of gravity of modern science is shifting more and more away from the sphere of the real world and closer toward that of possible worlds, in that it points the way to the realization of new phenomena not existing in the real world hitherto, not only expanding the horizons of the real world in width and depth but also pointing to the possible frontiers of such expansion, to the limits of possibility for the solution of tasks of a certain type, etc.

(b) Science as a cognitive activity *sui generis* leads to the formation of what we know as scientific data, methods and theories. Viewed in this way science appears as a *relatively consistent system of data*, or as a system of methods or general directions as to how to arrive at these data, or data of an analogous nature, or finally to data hitherto unknown. We might describe these aspects as the *methodological and theoretical feature of science*.

Data as the results of scientific activity are always such as to be communicable and therefore relayable in time and space. In other words, *science produces information* which aids man in his decision-making, is capable (at a given level) of optimizing this decision-making and of reducing the uncertainty in decision-making, whether, moreover, the decisions are to be made within the framework of science itself or in various domains of practice: manufacture, technology, public health, cultural life, etc. This also means that the results of scientific activity can be *applied*, i.e. they can reduce (or eliminate) uncertainty in decision-making and improve the quality of the latter.

(c) Activity in science is also connected with certain social institutions *(the institutional aspect of science)*. This aspect has always been very important and has always played a major role in the formation of the social relations and conditions of those involved in science, at universities for instance; nevertheless, the last few decades have shown a considerable strengthening of this aspect: in connection with the sharp upward trend of all the rough indices which characterize the scope of scientific activity (the number of persons actively engaged in science, the number of publications, costs of and investment in science, etc.), there has emerged, in almost every country, a mighty system of scientific bodies, institutions, etc. In the developed countries we find as much as one percent, or even more, of the total population engaged in what for this very reason has been described as 'big science'.

The emergence of big science has brought new sociological and psychological elements into scientific life; activity in science has become a profession in the purest sense of the word. It has also brought about the formation of a multilevel and, as a rule, hierarchically ordered system for decision-making about science with all its complex situations, which are sometimes even conflict situations.

(d) Science is also a social and socio-psychological phenomenon *(the sociological and psychological aspect of science)* which is having a continually greater influence on the other aspects of our social life. The achievements of science are penetrating into practically every domain of our existence, sharing at a steadily growing rate in the development of, and progress in, all these domains. This is why we sometimes hear of the scientization of our daily lives.

Once science is described as a social phenomenon, certain questions arise as to its place in the structure of the other social phenomena. If we accept the differentiation of social phenomena into various levels, for example phenomena of the [material and productive] foundations and those of the superstructure, then we may say that from this point of view modern science represents a typical multilevel complex which through some of its properties and manifestations belongs among the phenomena of the superstructure, and through others – and at an ever more conspicuous rate – comes into direct contact with the sphere of production forces. For this reason we cannot simply draw a dividing-line between that conception which describes science as one of the forms of social consciousness and that which emphasizes the importance of science as a factor operating directly in the progress of technology and production. The formulation that science is more and more becoming an immediate force of production is more by way of being an abbreviated statement to the effect that the state and level of production forces, of contemporary technology, and of course, first and foremost, of man himself, are increasingly dependent on the achievements of science, which does not, in itself, produce material goods (except for certain unique apparatus) but rational information which is of absolute importance for any further progress in the forces of production proper.

(e) A list of the major aspects connected with the concept of 'science' would not be complete without mention of the *ideological aspects* and ideological connections of science. The achievements of science cannot but influence what is traditionally known as *Weltanschauung* and fundamental ideological attitudes to society and to the position of man in the world and society. In the same way, no scientific activity can be absolutely cut off from ideological attitudes.

2. The goals of science as cognitive activity *sui generis*

If we consider science as conscious and organized cognitive activity, we must, in any assessment of this activity, take into account the broad spectrum of the goals it sets. From out of this broad spectrum of various and frequently quite heterogeneous goals, let us consider those which are directly connected with cognition, namely the epistemic goals. (This is not, of course, to underestimate the significance of other goals – social, psychological and so forth – the assessment of which is not a matter for the present work.)

It is sometimes said of science that its aim is to make predictions, or to make explanations of phenomena as a possible background for prediction. While such a definition of the goals of science is unquestionably very narrow, it does capture the quite substantial component of science as conscious and organized cognitive activity.

Scientific explanation is usually taken from the intuitive point of view to mean the kind of procedure which leads to the discovery or deduction of a scientific law or hypothesis which enables us to make sufficiently reliable *predictions*, or opt for such decisions the effect of which ensures the realization of desired aims with an adequate reliability. The explanation of a satellite's movement by the appropriate laws enables us to predict its future states with accuracy; an explanation of the cause or actual essence of cancer would enable us to decide in favour of such measures as would be likely to render its further occurrence impossible. We must then consider as a typical component in scientific explanation the discovery of such regularities, hypotheses or laws as enable us to make predictions or produce the requisite measures or precautions to achieve desired aims. In this sense we can also understand the abbreviated formulation that the goal of science is *prediction* or *production*. A subject for explanation can only be an event or class of phenomena or events, explanation being the discovery of suitable formulations to become a starting-point for prediction or production.

It would be looking at things too narrowly to confine the goals of science as cognitive activity merely to scientific explanation or to whatever might be connected with it, i.e., in the main, prediction and production. If these components of science as conscious and organized cognitive activity lead to certain discoveries, which in turn lead to certain formulations (sentences, hypotheses, laws, etc.), then the necessity arises of verifying the reliability of these formulations. Among this group of goals of scientific activity belong such procedures as verification, confirmation and testing. *Verification* is usually understood to mean a procedure capable of ensuring a final decision as to the truth of a formulation. Verification can be applied both to a datum expressed by a singular statement and to data expressed by general statements. In the latter case, of course, verification has its limits owing to the limited possibilities of verifying each individual possible instance. In contrast to verification, from which a final decision is expected, the assumption in the case of *confirmation* is that the statement will be confirmed to a certain degree, leaving open the possibility of subsequent steps which might eventually raise the level of confirmation. Confirmation may be visualized as a sequence of steps, e.g. experiments, measurements taken, etc., or of a particular ordering of these steps, which lead towards the gradual refinement of confirmation.

It goes without saying that these procedures are variously bound up with each other and combine together. It is, moreover, quite apparent that we must be able to decide about facts and events as well as about the statements, and indeed any other formulations, by which we express the results of cognitive activity. This means that in scientific activity we combine together what may be characterized as *ontological decision* with what may described as *semantic decision.* Ontological decision relates to the facts and events of a given universe of discourse (henceforth merely 'universe') and indeed to any entities accepted as elements of that universe, i.e. things, properties and relations. These entities can be *observed, measured, explained, predicted,* etc. On the other hand semantic decision relates to statements or any other expressions of language capable of expressing the results of cognitive activity and which can be considered as relatively independent expressions of the language which we use with respect to the given universe.

The differentiation between ontological and semantic decisions in assessments of the nature of scientific procedures is not absolute. This differentiation must be relativized according to the relation between the given universe (and those entities accepted as elements of that universe) and the language, and the relatively independent expressions of the language, which we use with respect to the given universe.[1] In the case of observation and measurement, it is usually assumed that they relate to entities of a given universe. It is, of course, perfectly possible to observe and measure written or uttered statements as well, but in this case the statements become an object of cognition and form a component part of that universe with which our cognitive activity happens to be dealing. It is customary in such cases to speak of statements in the object-language as distinct from statements in a metalanguage.

Unlike observation, measurement, explanation, prediction and other procedures which we relate to the entities of a given universe, explication, interpretation, definition and certain other similar procedures do not relate to the entities of that universe but to statements or any other expressions used with respect to the given universe. Here we are dealing with the delimitation or refinement of the meaning of these expressions, the replacement of expressions with less precise meanings by expressions with more precise meanings, and so forth. These then are procedures connected with semantic decision.

This review of the groups of scientific procedures which also comprise the relative goals of cognitive activity in science is of course neither complete nor exhaustive. From this review it follows that these scientific procedures are connected either with ontological or with semantic decisions. It also follows that the epistemic goals of science cannot be connected with ontological decisions alone. In other words procedures which are carried out in science cannot be reduced to merely ontological decision-making.

There is another important reason for pointing out the difference between ontological and semantic decision. In any analysis of the goals of cognitive activity in science or in the analysis of scientific procedures, sight should never be lost of the *language* in which we formulate the results of these procedures relating to entities in the universe under investigation. Any results arising out of those procedures which have the character of ontological decisions, are communicable only in as far as they are expressed by means of expressions from the language which we use with respect to the given universe. These expressions may then become the object of various forms of semantic decision. Connected with the language of science are certain other scientific procedures concerning the delimitation of the vocabulary, the establishment of the initial elements of the vocabulary, what has been called the *formation of concepts*, *constitution*, *reduction*, semantic and syntactic relations of the given language.

Cognitive activity in science is a complex and heterogeneous set of procedures which, according to the nature of the immediate task, come sometimes more, sometimes less to the fore. This applies not only to the various different segments of activity in science but also to the different stages in the solution of a task. These procedures may also be considered as relative or temporary goals in science, it being obviously true that the achievement of more significant goals presupposes that of other goals which are the prerequisite for the more significant goals.

The differentiation of scientific procedures from the point of view of ontological and semantic decision serves our purposes only as long as we are considering the epistemic goals of activity in science. However, we have already indicated that science should be understood also as a system of methods or general directions for how to arrive at new data. Taking into account the methodological aspect of science, we must also pay due attention to the fact that scientific procedures can be refined, improved and replaced by others, and that it is desirable to optimize (or economize, etc.) certain procedures. In this sense we might speak of *methodological decision-making*. If we relate ontological decisions to the entities of a given universe and semantic decisions to statements or any other expressions of the language we use with respect to that universe, then methodological decisions relate to our own procedures as connected with the epistemic goals of scientific activity, or to individual components of these procedures. We have in mind here, for example, the selection of such procedures and their components as are capable of guaranteeing the attainment of goals at the level desired, the optimization of the respective procedures, decisions as to the degree of relevance of the individual components, e.g. the various parameters, the permissible degree to which certain components may be reduced, etc.

3. Scientific Cognition as the Solution of Problems

The concept of 'scientific discovery' (cognition) and 'cognitive activity in science' can be explicated not only from the point of view of goals (we have already outlined some of the more important ones), but also from the point of view of certain other features. We shall point here to some of the more important features which figure particularly in the connection between the concept of 'scientific cognition' and such concepts as that of 'problem', 'solution of a task', 'uncertainty' and 'operation with data'. Thus scientific cognition may also be characterized as (a) the solution of problems, (b) the solution of tasks, (c) the elimination of uncertainty, or (d) operations with data. We cannot, however, overlook the fact that all these characterizations present a sort of *genus proximum* with no specific differentiation, which is therefore quite broad and involves other phenomena which cannot be themselves considered as elements of cognitive activity in science. Nevertheless these viewpoints may be of use in affording a broader view of cognitive activity in science. In what is to follow we shall be relying on the viewpoint stated and on the allied standpoint which might be characterized as the communication model of science.

(a) In ordinary language the concept of 'problem' is used generally in a way that is very vague and far from unambiguous. If we describe cognitive activity as the solution of problems [2] (or rather the solution of problems *sui generis*, as connected with a certain system of goals and procedures as outlined in the foregoing section), then a certain explication must be given. We speak of a problem, or problem-solving situation, if there is something unknown which is to come to be known, something undecided which is to be decided, something which is an obstacle to activity and is to be overcome, etc. One important thing is that the problem is not just anything unknown, but *something* unknown, undiscovered, undecided, something which we need to know, discover, decide, overcome, etc. Accordingly, a problem is not simply any unanswered question, but one component of a problem is the question which for one reason or another we want, need or have to answer. It is evident that although a component of any problem-solving situation is something unknown, undiscovered, undecided etc., we are still capable of getting to know and characterizing this something. This means that not only questions but also certain statements are a component in the expression of problem-solving situations. These statements may, moreover, be of the various kinds which combine with questions.

The connection between these two components in the expression of a problem-solving situation – with respect to the goals typical for cognitive activity in science – can best be demonstrated by a few examples, which we give here in the form of a simple diagram (see Table I).

It could be shown by similar means that a certain combination of statements and questions is also characteristic for the expression of other specifically scientific problem-solving situations. Analogous combinations of statements and questions probably also occur in the expression of problems outside the sphere of science. (We speak, for instance, of moral, economic, production problems, and so forth.)

(b) The concept of 'solving a task', as a concept capable of throwing light on some aspects of cognitive activity, is also broader in scope than

TABLE I

Statements	Questions	Goal of solving the problem
Statements of an empirical nature describing the course of a given process	What is the explanation of the phenomena described?	Scientific explanation
Statements as to the regularity of the development of a given system	What will be the state of the given system at time t_n if at time t_0 its state is S_0?	Prediction, retrodiction
Statements of hypothesis h	In what degree is hypothesis h confirmed by evidence e?	Confirmation
Statement of a synthetic nature	Is a given statement true?	Verification

the concept of scientific cognition and covers many activities quite distinct from science. Tasks may be solved in production, party-games and several other spheres of social activity.

There would be little point in attempting a general definition of the concept of 'task' which would be adequate to every possible instance. The explication of this, intuitive, concept depends on concrete circumstances, means and goals. It would be more appropriate to attempt an explication of the concept of 'solving a task'. The concept of 'solving a given task' is usually understood as the reduction of the solution to a finite sequence of tasks which we take as solved. If our task is to prepare Welsh rabbit, we must know what we need for the purpose (bread, cheese, mustard), where it is to be obtained, what operations we must perform and in what order we must perform them. At the same time we assume that each of these intermediary steps represents a task the solution of which is known. This analytical approach to the solution is by no means the inevitable rule. If we know where to buy food ready-made, the solution of one task can be carried over to that of another. A similar state of affairs can be found with other tasks: if we know how to add and subtract and are capable of reducing multiplication to addition, then we can reduce a task involving multiplication to a sequence of tasks using addition. If we know how to multiply, then we can reduce raising to a power to a sequence of multiplications.

What we have said so far in explication of the concept of 'solving a task' goes to indicate that this concept cannot be understood in any absolute sense. The explication of the concept of 'solving a task' (or rather 'solving a given task') must therefore be relativized to the collection of means available for the solution of the given task, or means which we must find for the purpose from among the collection of all known means, or even means which are hitherto unknown and must therefore be discovered first. To take an example of a simple task: We are to construct an isosceles triangle having two given altitudes. Of course we need to know what means are available. Generally in such cases it is straight-edge and compasses. With these it is possible to construct an isosceles triangle with two given altitudes in several ways. (For example, one can use the ratios of the sides and altitudes and construct the angles; or one can erect one altitude on a straight line and draw the tangents from its vertex to the circle with centre at the foot of the altitude and diameter equal to the second altitude.) Thus, when deciding upon the approach to the solution of a given task we take into account the means which we have at our disposal. In this case general instructions can be formulated which respect the means given, and which may be considered effective for the solution of the task provided an effective solution is found for the intermediary tasks to which the given task has been reduced.[3]

This intuitive explication of the concept of 'solving a task' should also have displayed the connection of the concept with the intuitive concept of algorithm. If we have a series of tasks of the same type, and if we can give a precisely determined prescription for the solution of the task in the shape of a finite sequence of individual steps (intermediary tasks), then we say that the task is solvable algorithmically. It goes without saying that not all tasks are solvable algorithmically, and that when they are then the algorithm is known beforehand.

If we say that scientific cognition is also the solution of tasks (tasks *sui generis*, that is), this does not mean that all scientifically relevant tasks are solvable algorithmically. The connections between the concepts of 'scientific cognition' and algorithm are more complicated than this. We can in fact say that cognitive activity in science is of an algorithmic nature only in part. When looking for new solutions we quite naturally make use of those already known, and in the search for new paths to follow we employ to the fullest everything that has been discovered previously. Thus there are both algorithmic and nonalgorithmic elements in scientific cognitive activity.

From the point of view of algorithmic solvability, we find in scientific activity tasks which are algorithmically solvable and have been solved, tasks which are algorithmically solvable but still awaiting a solution, tasks which are not solvable algorithmically but which can be solved by other means (such as an individual task solved by trial and error), and of course tasks which are definitely not solvable but whose unsolvability must be demonstrated. For this reason a quite considerable portion of time and effort in science is devoted to looking for new algorithms and also for ways of proving that a given algorithm is impossible. Nor can we eliminate from scientific activity, at least, that is, for a critically limited period, endeavours to solve those tasks which are definitely unsolvable. It is true that examples of such endeavour are generally adduced from the prehistory of science (the search for the elixir of life, *perpetuum mobile*, etc.), nevertheless not even in modern science can we exclude certain steps which only in time turn out to be as futile as the search for the *perpetuum mobile*.

Thus from the point of view of the concept of 'algorithm' it is not only of importance that scientific cognition leads to the discovery of previously unknown algorithms but also that it shows when certain algorithms are not possible. In other words the discovery of how to solve certain tasks is only one of the valuable results of scientific cognition, another being the discovery of the limitations of such solutions or the discovery that a given task is in fact unsolvable.

When considering the connections between the concepts of 'scientific cognition' and 'algorithm' we also see clearly that science cannot be visualized as the magic power capable of solving everything, which modern society has turned into a fetish to replace the imaginary supernatural powers of earlier beliefs. The spirit of modern science has no share in this fetishism nor in the ideas of the omnipotence of science or ideas of a scientific fideism or religiosity. The thesis that 'science solves everything' no longer holds and remains equally primitive and dubious even if we replace the term 'science' by other terms denoting abstract or concrete objects.

(c) The link between the concepts of 'scientific cognition' and 'uncertainty', or rather of the 'elimination of uncertainty', has come about thanks to the scientific results brought by cybernetics and some of the disciplines allied to it. When we characterize scientific cognitive activity as endeavour leading towards the reduction or elimination of uncertainty, we must once more bear in mind that this *genus proximum* is again wider in scope than the concept of scientific cognition as such, and that it also involves certain other processes which are not in general considered as scientific cognition, in particular various kinds of communication and control.

Scientific cognition is of course a *major component in the negentropic activity* of man, i.e. his effort to strengthen what Norbert Wiener (Wiener, 1954) has called 'local enclaves whose directions seem opposed to that of the universe' (p. 20), i.e. directions towards strengthening the measure of organization and reducing the measure of uncertainty. Through his scientific activity and the application of its achievements in practice man overcomes – albeit only in certain segments of nature – the general statistical tendency of nature towards increasing disorganization. Thus scientific cognitive activity may be described as a specific highly rational form of the resolution of conflicts between man and his changing, only partially known, environment. Norbert Wiener expresses it as follows: "The scientist is always working to discover the order and organization of the universe, and is thus playing a game against the arch enemy, disorganization." (p. 50)

In recent times Léon Brillouin has been one of the main persons to draw attention to the connection between the concept of scientific cognition and the reduction or elimination of uncertainty. He emphasizes that a 'scientific law' represents a method of prediction enabling us to build up highly negentropic systems (Brillouin, 1964; p. 18). Although he understands the concept of 'uncertainty' primarily in the statistico-physical sense, it is quite apparent that it also has important semantic and pragmatic connotations. In other words, scientific cognition is driven on the one hand by the effort to reduce uncertainty in nature itself and so increase man's control of nature (by way of prediction or production of certain phenomena or processes), but also on the other hand by the effort to reduce the uncertainty of man himself, uncertainty arising out of his state of weakness, restrictions or ignorance.

The concept of the 'elimination of uncertainty' is yet another concept which cannot be understood in the absolute sense but only with respect to certain tasks or problems of decision-making. When we say that science is an endeavour tending toward the reduction or elimination of uncertainty, there is a further converse principle which also holds, namely that science itself produces uncertainties and is the source of new unanswered questions, new problems and new difficulties. With Wiener emphasizing the homeostatic function of science, i.e. a function tending towards strengthening man's position in his conflicts with his changing and only partially known environment, we cannot fail to see that this function is in fact ambivalent. All new discoveries or new solutions of known problems may give rise to new problems and naturally to new difficulties and dangers, which in turn call for further new discoveries. These difficulties and the increase in danger are the price man pays for the labour and effort spent on the way to scientific progress. The ambivalence of science is symbolized by the figure of Prometheus whose great deed brought untold benefit to mankind but only at the cost of immeasurable distress. It is true not only of the discovery of fire, but also of many other major discoveries which were or are the results of scientific endeavours, that they broaden man's potential, eliminate uncertainty, and reduce the risk element in human decision-making, while at the same time giving rise to uncertainties in other connections and creating new and unexpected risks.

NOTES

[1] In the account to follow we shall be attempting to give a more precise description of this relative independence in connection with what we shall characterize as the communication model.

[2] Some stimulating ideas on this approach to the concept of scientific discovery have been expressed by Polikarov (1967) who draws the distinction between two aspects of scientific cognition: the static aspect, which involves the construction of a system out of already existing data, and the dynamic aspect, which he understands as a process in which he distinguishes three stages: the problem, projects for its solution, and the selection of the optimal solution.

[3] The concept of 'solving a task' is not very easy to distinguish from that of 'solving a problem'. In fact the two terms considerably overlap, at least in their everyday intuitive use. However, when speaking of 'solving a task' we do usually understand that the goals of the solution have been explicitly, or at least implicitly, set, as well as the set of means which we have at our disposal for the purpose.

CHAPTER II

ON THE APPROACH TO MODELS OF SCIENTIFIC PROCEDURES

1. Scientific procedures as operations with data

So far we have been speaking of scientific explanation, prediction, verification, confirmation, constitution,[1] reduction, etc., as goals of cognitive activity in science. It is evident that any deliberate activity, i.e. any conscious and rationally organized sequence of operations resulting in the attainment of the appropriate goal, may also be described as explanation, prediction, reduction or constitution. Accordingly, wherever we consider *scientific procedures* in what follows, what we shall have in mind will be activity in this sense of a conscious and rationally organized sequence of definite operations.[2]

It should also be apparent from the foregoing that in so far as we consider scientific procedures as elements of scientific cognitive activity, we can characterize and analyze them as the solution of problems and tasks and as an activity leading toward the reduction or elimination of uncertainty, or as *operations with data.*

In what is to follow we shall be paying most attention to the last of these approaches, that is we shall be considering scientific procedures as operations with data, although even this *genus proximum* is much broader in scope and includes other operations than those which are actually components of scientific activity. Two fundamental questions arise in this connection:

(a) What is the nature of scientifically relevant data?

(b) How are we to understand the concept of 'operation' in the analysis of explanation, prediction, constitution, reduction and other scientific procedures?

We shall first try to show that no answers to these questions can be given without the prior acceptance of certain important assumptions. While it is true that in general a distinction is drawn between empirical data (e.g. the results of observation, measuring and experimentation) and theoretical data (e.g. scientific hypotheses, scientific laws), this is no more than a very rough and scarcely satisfactory division. Likewise, as concerns the concept of 'operation', its interpretation depends both on the nature of the data and on the approach to the presentation of the operations. A distinction is sometimes drawn between empirical procedures (which include, in particular, observation, measurement and experiment) and theoretical procedures (e.g. explanation, constitution, reduction, proof), the main attention being paid to the nature of the data concerned in the respective procedures. This differentiation is, like the previous one, very rough. Even were it possible to discriminate between empirical and theoretical data, it remains apparent that a great many scientific procedures presuppose operation with both theoretical and empirical data. In the case of prediction, for instance, due account must be taken of empirical data relating to the states of the system under consideration as well as of knowledge of the relevant laws.

In any explication of the concept of 'operation' it is of particular importance what system we are going to use, how the data are represented, and what methodological devices are selected. Some procedures can be represented as the kind of deductive method where statements of one kind are derived from statements of another kind: in the case of diagnosis, for example, we assume that we have at our disposal certain statements which fix the general connections between kinds of indisposition and symptoms, and statements on the symptoms discerned in the patient. From these data we then try to deduce the data to express the presumed diagnosis. Viewed in this way, the concept of 'operation with data' ('data' being understood to mean that which is expressed in the statements alluded to) is explicated as a logico-deductive procedure whereby the conclusion containing the disorder we have sought to identify can be derived from premisses containing statements on the symptoms found in the patient and statements on the general connections between the symptoms and the disorder. However, we might also imagine a different system in which the explication of the concept of 'operation with data' and, in this case, the diagnostic procedure would take on a different form. This procedure may also be treated as a proces of decision-making, the operation which leads to the establishment of the diagnosis being then understood as the selection of a certain decision from among a set of possible decisions on the basis of established assumptions and known decision functions.

Let us now try to summarize some of the characteristic features of scientific procedures as operations with data. The main ones are as follows:

(a) Any scientific procedure which we consider as an operation with data (more precisely, an operation with data *sui generis*) is *goal-seeking*. This is connected with the fact that the results of the various scientific procedures, i.e. explanation, prediction, diagnosis, etc., may be considered as relative goals of scientific cognitive activity, as we have already indicated. The goal-seeking character of scientific procedures generally also has certain other mediated connections: although the scientific explanation of a process under investigation may be a relatively immediate goal of the investigation itself, this activity usually has still other goals; it is not merely explanation for explanation's sake, but scientific explanation with the aim of optimal production, the most accurate predictions and so forth. And similarly with diagnosis: while the immediate aim is to diagnose the patient's case accurately, the actual diagnosis is generally intended to serve in the selection of appropriate treatment. Also in the case of the procedures usually considered as empirical, e.g. measurement, experiment, etc., distinctions can be drawn between the immediate goals – for instance the quantification of certain properties – and certain other goals.

Taking then scientific procedures as operations with data, their immediate goal is to establish certain resultant data on the basis of data which we already have at our disposal, or on the basis of data which may be acquired additionally in the process of the given procedure. Moreover, these resultant data usually serve certain further goals: the measurement of co-ordinates used to serve ships in finding their location and in navigation, diagnoses serve in the selection of an appropriate cure, and so forth.

(b) All scientific procedures are characterized by what we might call *relative apriorism*. It is desired to discover certain data, but only on the basis of data which we already have at our disposal; it is desired to make a decision on something, but this 'something' must be stated beforehand. Figuratively speaking, scientific cognition is not generally a penetration into an entirely unknown world, but into a world which is in fact partially known and which we are able to characterize up to a point (albeit without absolute accuracy, and sometimes even quite erroneously), or about which we have certain hypotheses. The relative apriorism may rest not only on our theoretical or empirical knowledge to date but also on certain erroneous hypotheses which will have in time to be cast off with the development of the given scientific procedure. It is familiarly known now that the hypothesis of an ether wind was erroneous; nevertheless it was an important component of the assumptions behind the Michelson-Morley experiment.

(c) Another important feature of scientific procedures is what we might describe as the necessity of the *finitistic approach*. Considering a scientific procedure as an operation with data, we might imagine the operation as a *finite* sequence of certain intermediary steps (operations, tasks), each one of which can be realized.

There are of course a variety of ways in which the finitistic approach can be realized. In systems within the framework of which the explication of scientific procedures as operations with data is carried out, the finitistic approach may be realized by our taking as finite some or all the classes of elements which constitute the given system. We shall be returning in due course to some of the possibilities in the realization of the finitistic approach.

The standpoint from which we characterize scientific procedures as operations with data might be designated as a communicational conception of scientific procedures, or more generally as the *communication model of science*. This standpoint, which lies behind the analyses to follow and which will be given further specification in the course of those analyses, naturally does not claim to be the only, all-embracing, explication of the concept of scientific cognition and the epistemic functions of science. Other possible standpoints are doubtless equally legitimate. Two circumstances have been the prime movers behind the selection of the communication conception. First, it should be noted that the generality of the conception is sufficiently reliable. Although the whole of what we may describe as cognitive activity in science gives a very heterogeneous spectrum of procedures which are highly different in form, largely according to the nature of the respective scientific disciplines, it remains self-evident that this activity *produces useful information* in all the branches of science. [The usefulness of the information is of course measured by quite different criteria in the natural sciences, the social, technical or medical sciences, as well as in those areas of productive, economic, technical or social practice which make use of the achievements of these sciences.] The other advantage of the communicational conception is that it lends itself fairly readily to the expression of those characteristic features of scientific procedures which we have already pointed out, namely their goal-seeking character, the role of relative apriorism, and all that follows from the requirements of finitism.

2. The metatheoretical character of the analysis of scientific procedures

In the explication of the concept of 'scientific cognition' and in the analysis of scientific procedures, it is essential to respect the meta-theoretical character of these problems. This means that statements made in this field differ in level from the ordinary statements of the science whose procedures are the object of investigation. Every science relates its statements to a certain 'universe of discourse'; this universe may include not only phenomena which are empirically or experimentally recordable (such as natural phenomena, social phenomena, the behaviour of natural and social systems, and so forth), but also phenomena which are the result of conscious human productional, technical or intellectual activity (for example, technical systems, systems of norms, systems of mathematical or logical objects, and so on). Thus the ordinary statements of science relate to the objects of some such universe. On the other hand, statements concerning scientific procedures do not relate to the same objects but to procedures the constructive elements of which are the very statements of science. This in fact means that the analysis of scientific procedures is a system of statements relating, directly or indirectly (indirectly as the constructive elements of scientific procedures), to the statements of science. In other words, the language in which we express the analysis of scientific procedures is a metalanguage with respect to the object language in which science expresses the results of its cognitive activity.

Respecting the metatheoretical character of the analysis of scientific procedures also places the very strict requirement of discriminating between *what* science makes its statements *about* – which is realized in statements about the objects of the given universe, their properties, relations, and the interconnections among these objects and relations – and *what* it *actually* states (which may be characterized by the use of statements about the former statements). The emphasis on distinguishing these levels may be very beneficial in the development and solution of the problems of scientific methodology since it enables us to overcome some of the illusions of *Naturphilosophie* which have arisen out of the confusion of knowledge of, or opinions about, the objects of a given universe, with opinions about these opinions.

The manner of verifying statements at the two levels is also quite different. In the case of the actual statements of science, the choice of an expedient and productive form of verification depends on the nature of the given universe: if this is a system of empirically apprehensible objects, the verificatory procedures must rest on such criteria (e.g. the operations of measurement, experimental procedures) as make adequate decisions possible about given objects, their properties and relations. In the case of a system of abstract objects (such as those we encounter in mathematics and logic), the verification of the individual statements about the objects of the given system takes on a different form. [It might, for instance, take the form of establishing the deducibility of a given statement from the axioms accepted and on the basis of the rules introduced.] The verification of statements relating to the statements of science, i.e. the verification of statements in the metalanguage, must also respect the nature of the objects of the respective universe: these, however, are the actual statements of science, or individual components of them.

3. The finitistic approach

In the analysis of scientific procedures, as we have already remarked, it is expedient to respect what we have described as the finitistic approach. We shall now try to demonstrate that finitism here cannot be understood ontologically, but methodologically, for which reason we shall speak of *methodological finitism*. Intuitively, the main thing here is that we should be able to carry out the solution of a given task in a finite number of steps (or elementary operations), in a finite time and with a limited range of means at our disposal. This means, for example, that the protocol of the solution must be set forth using a finite number of formulae. If the task cannot be carried out in a finite number of steps, or if its solution requires an infinity of time, then it cannot be deemed solvable.

It might appear that the requirements of methodological finitism are quite naturally substantiated. Nevertheless, we shall try to justify the grounds for them more fully, pointing at the same time to certain important methodological consequences ensuing from these requirements. If we are to consider scientific procedures as operations with data (*sui generis*), it is essential to respect the fact that these operations are realized within systems having, from the communicatory point of view, only finite 'capacities', 'memories', and 'delays' at their disposal. Man himself is naturally one such system, although as far as future potential is concerned he may continually expand and improve those of his properties which correspond to the 'capacities', 'memories' and 'delays' of communication systems. Any particular improvement or expansion is of course in itself only finite and hence changes nothing in these characterizations.

This justification of methodological finitism may be supported by certain further arguments based on the theory of levels.[3] This conception, which is based on the idea of the 'qualitative infiniteness' of nature and an infinite number of levels, stresses that at each level we operate with a finite number of parameters which are fundamental to the given level. Hence no scientific theory which is linked with only one or a very few levels can give a picture of the totality of nature. For the same reason any scientific theory, if it is to be of practical use in decision-making, is based not only on a partial view of nature, but also on the finite number of parameters which are fundamental to that view. With the transition from one level to another, the horizon of these parameters naturally changes, new (at the initial level 'hidden') parameters emerge while others lose their previous significance. Hence, in physics for example, the transition to the 'lower' levels leads to the loss of significance of the original macroscopic concept of 'heat'. On the other hand, we know that at the quantum level such concepts as that of 'spin' are to be found which do not occur at the 'higher' levels.

For the justification of methodological finitism we must always have in mind how essential it is to take account of only a finite number of parameters for the operations which can be accomplished at each separate level. The problems of reduction and constitution, which present themselves here as problems of narrowing-down or expanding the number of parameters of which account must be taken, are then closely associated with transition from one level to another.

The conception of the theory of levels can be connected to that which we have described as the communication model of science. Such an alliance may moreover contribute to the elucidation of the rather vague concept of 'level'. The communication model assumes that we always approach a certain universe armed with certain (relatively limited) equipment represented by the potentials of our measuring instruments, our experimental techniques, and the theoretical means we apply. No analysis of a domain of objects is at all possible without these means, which always have a finite degree of discriminatory ability, finite thresholds (i.e. the limits beyond which the discriminatory ability of these means ceases to function), and so forth. Also there are similar limitations to the ways in which the set of means used with respect to a given universe can react to the stimuli coming out of this universe, and record and transform the stimuli.

Some objections to methodological finitism might arise out of the misunderstanding caused by the incorrect substitution of methodological finitism for ontological finitism. It should be stressed that the former can under no circumstances be identified with the latter, i.e. the idea of the finite and restricted nature of the objective world. Methodological finitism does not therefore conflict with the viewpoint that the objective world is infinite and inexhaustible. The core of the matter is merely that our means of analyzing the world (and a major role in the breakdown of these means is played by scientific procedures, the subject of this book) are at any given moment finite and restricted. It further goes without saying that it is not only up to a point possible, but also quite essential, to expand, perfect and elucidate these means. Hence the conception of methodological finitism simply cannot be interpreted in the sense that the finite and restricted nature of our methodological means is absolute and invariable.

NOTES

[1] We are using the term 'constitution' in the sense of German 'Aufbau', i.e. as an operation inverse to a reduction, e.g. the introduction of new terms and expressions etc.

[2] The fact that so far we have been using the same set of terms for both a scientific procedure and its result, need not be particularly disturbing provided it is always clear which is concerned in any given instance. In some cases the two are distinguished: by measuring we understand a sequence of operations, the term 'measuring' not applying to the result. In other cases the distinction between a process on the one hand and the result of the process on the other hand, may be made by a convention: for example, the convention might be introduced

whereby the term 'abstraction' denotes activity, and the term 'abstractum' the result of this activity.

[3] The theory of levels is a philosophico-physical conception which arose out of certain methodological difficulties created by quantum physics and some other parts of modern physics. The main representatives of this conception are generally taken to be Bohm (1957) and Vigier (1962).

CHAPTER III

THE EMPIRICAL BASIS AND THE ANALYSIS OF THE 'UNIVERSE OF DISCOURSE'

1. Schemata of the Analysis

(a) *External Specification of the Universe*

It is an obvious assumption of scientific activity that its results, i.e. statements of science, relate to something. In other words we assume that statements of science refer to a certain sphere of objects, the totality of which we have called a 'universe' (universe of discourse). We do not of course mean by this the universe of all sciences as a whole throughout their historical evolution. When speaking of a universe, what we shall have in mind will be a particular sphere which is in some way or other specified, and which, other things apart, is also relativized to a certain sphere of science, the latter being at a given stage in its development. It follows from this that universes so understood must of necessity be considered as merely components of the objective world. This latter remark must also be taken as meaning that a universe may be enlarged by the inclusion of new objects for which the original definition made no allowance.

From the formal point of view a universe is a certain non-empty set of objects which are in some way or another characterized. Hence we also need some criterion which enables us to decide of any given object whether or not it is an element of that universe and so whether we are to take it into account when analyzing the universe. The criteria governing this kind of decision-making in science may be of the most various nature: for the empirical and experimental sciences they will be, in the main, empirical and experimental criteria. We take it as a matter of course that biology deals with the phenomena of *living* nature, that thermodynamics studies phenomena of *heat*, acoustics those of *sound*, and so on. It might indeed be objected that 'living', 'heat' and 'sound' can be treated as empirical ideas only in everyday speech and that in contemporary science these and other similar ideas can by no means be considered as purely empirical nor even as distinguishable through human sensory activity.

However, this kind of objection is completely without foundation. In the development of science we are witnesses to how many originally purely empirical criteria change in significance as the universes of the individual branches of science become specified, and of how an ever increasing measure of attention is paid to theoretical criteria. Moreover, the majority of the criteria which originally functioned as empirical are treated in modern science as theoretical criteria. And nowadays many theoretical criteria are considerably removed from the original purely empirical criteria. To take an example: If we define the sphere of our investigation by saying that we are interested in 'phenomena of linear oscillation', we need not concern ourselves first and foremost with what kind of oscillation, whether electromagnetic, mechanical or some other kind; on the contrary the first thing which should interest us are the theoretical characteristics which are common, empirically, to these quite heterogeneous domains. Similar remarks might be made to apply to processes of control, information, etc., since here too the focus of attention is the structure of these processes, their adequate expression in the appropriate mathematical formal apparatus, but not the empirically heterogeneous aspects of the systems within which the processes of control and communication are applied.

The criteria for specifying a universe may also be of a conventional nature. If we study the properties and behaviour of systems of abstract constructive elements, such as Turing machines (which are not, and in principle cannot be realized technically and so not even empirically), it is expedient to establish beforehand, and on the basis of some convention, what objects are to be included in the respective universe.

In science there is simply no question of our being able to establish any kind of priority of certain criteria over others in the specification of a given universe. Although we are witness to the fact that the selection of these criteria is increasingly dependent on theoretical data and theoretical goals, we cannot overlook the presence of also pragmatic considerations made necessary primarily by the purpose which the given investigation serves or is intended to serve. We often find in science cases of mutual interference between what is called the 'object of science' and what comprises the set of methods, data, theoretical and experimental means of that science, including the goals whose achievement these means serve. The choice of universe naturally affects the selection of methods, theoretical and experimental means, etc., and on the other hand this set of methods and means has a considerable effect on the specification of the universe.

Very illustrative of these processes and so too of the gradual changes in the specification of the universe, is the evolution of some of the branches of physics. To take the example of the origins and development of acoustics: the first stage was the study of the phenomena perceivable to human hearing, and the first steps in the specification of the universe were evidently based on purely empirical criteria. Subsequent study of acoustic phenomena, the study of vibration, the discovery of the Doppler principle and other advances in the discipline, led not only to the inclusion in the original universe of phenomena which human hearing is simply not adequate to distinguish, but also to fundamental changes in the criteria applied in the specification of acoustic phenomena.

(b) *Internal Analysis of the Universe*

Our only considerations on the universe of science up to this point have merely been to indicate some of the problems involved in its external specification. However, the problems associated with the internal analysis of the universe are of no less importance for the methodology of science. It is quite natural that this internal analysis depends very largely on the very nature of the universe itself. In the account to follow we shall have a look at some of the more important schemata of analysis, in particular as applicable in the empirical and experimental sciences.[1]

One basic fact to be pointed out is that some of the fundamental schemata of the internal analysis of a universe which we encounter in the empirical and experimental sciences, have arisen out of common sense and are based on certain categories of everyday speech. This is the case particularly with the schema 'thing – property' (or the expanded schema 'thing – property – relation'), yet the same applies, after all, to other schemata, that of 'situations' in the universe, and that of 'events'. Let us now take a more detailed look at the various schemata and at some of their logical and methodological connections. They are of course kept strictly distinct merely in view of the need for a more detailed analysis; in everyday speech and thinking these schemata are mutually interwoven and complementary.

(a) One of the most common schemata of the internal analysis of a universe is that of 'thing – property', or again its expanded form of 'thing – property – relation'. Let it be stressed in advance that it is not our intention here to analyze the traditional ontological, and in philosophical literature not very lucid, set of problems connected with these concepts. Instead, we shall be concentrating only on those questions which are connected with the analysis of a universe and the structure of the language of science.

The schema 'thing – property' is rooted in everyday speech and is supported by the most elementary version of empiricism: we are capable of distinguishing among the individual 'things' around us by way of telling apart the 'properties' of these things, the simplest version of the act of telling apart being sensory. The familiar dispute as to the priority of thing or property does not of course yet present itself at this primitive level. The dispute arises as soon as we begin to ascribe more profound philosophical and in particular ontological connections to the schema.[2] If we concentrate our attention on the formal aspects of the given schema and confine ourselves to the most general qualification of a 'thing' as an object (individual object) in a given universe, then it is true of any such thing that we can make statements about it. We usually interpret this ability to 'make statements' about a thing in the sense that we assign certain properties to it. Relation may also be understood as a property, namely that of a pair, triplet, ... *n*-ple of things organized in a particular way. Hence we shall interpret the expression 'to make statements' about things in the sense that we assign properties to the things, or pairs, triplets, ... *n*-ples of them, the term 'properties' being understood to include relations.

The circumstance that we are able to make statements about things (pairs, triplets, ... *n*-ples of things respectively) ordered in a particular way, will be called a *fact*. It should be emphasized that this conception of what we call a fact assumes the application of a particular schema of analysis of the universe, i.e. in this case the schema 'thing –property', or 'thing – property – relation' respectively. From this point of view it is naturally quite unthinkable to consider the existence of things to which no properties or relations belong, or to imagine properties and relations which are unassignable to things. If we were to accept a different schema of analysis of a universe, we should have to adopt a different conception of what we call a fact.

(b) Another schema of the internal analysis of a universe is that which is best described as the *schema of 'situations'*, or the schema of 'time-space domains', 'fields', etc. With the schema 'thing – property – relation', the analysis of the universe enables us to reach down to the individual objects, which leads to the assumption either that certain principles of individualization are available, or that the individual objects are themselves the starting-point of the specification, internal analysis and decomposition of the whole universe. In the case of the schema of 'situations' there is no need to assume that the analysis of the universe will lead us down to individual objects.

In answer to the question 'what is the situation' (which is generally understood to mean 'what is the situation in a certain assumed or defined space'), we usually say, in everyday speech, something by which we in fact characterize the distribution of certain parts of the universe, their interrelations and so on. A meteorologist answering this question (where what is understood is the 'meteorological situation' in a certain defined area) will say that in this part of the country it is raining, another is overcast, and in yet another sun is recorded. A staff officer answering the question as a query about the situation on the battlefield, will describe the distribution of his own forces, where the enemy has his tanks and infantry, and so on.

It might seem that the schema of 'situations' is a kind of modification of the schema based on the relations between the categories of 'part' and 'whole'. However, any such assertion is only partly justified, namely in as far as 'whole' is taken to mean the whole universe. It is in fact typical in the case of the schema of 'situations', that we take account of the relation of a given domain or field to the whole universe.

Although there is no need, as we have mentioned, to assume that the analysis of the universe in the case of the schema of 'situations' be carried right down to the individual objects, there is also no need to exclude this possibility outright. This is because we may always allow for the possibility that individual domains or fields also consist of single objects, although we are naturally not always able to distinguish these individual objects completely. The only essential matter here is that they may be considered as elements of a given class.

It is convenient in this connection to distinguish two degrees of individualization:

(1) The elementary degree of individualization occurs when we are able to demonstrate that a given object belongs to a given class or not.

(2) The higher degree of individualization occurs when we succeed in demonstrating whether or not a given object belongs to all of the classes which we are capable of distinguishing by the decomposition of the universe.

These same two degrees of individualization are also expressible by the means afforded by the schema 'thing – property – relation'. In the first case we are in a position to demonstrate only one property of the object. In the second we can determine all (known, accessible) properties of the object.

From the methodological point of view it is of great importance to distinguish the two degrees of individualization.

An example of the first degree is the discovery in diagnosis that the patient has influenza. Even this discovery is in itself of practical importance and can influence the measures to be taken, such as the choice of a cure. However, it is too bare for any more thorough examination of the patient's state or for making better decisions as to the adequate and effective treatment. For these purposes it would be necessary to pass to a higher degree of individualization and see what other properties the object has, or, which is merely another way of putting the same thing, see to what other classes the object belongs. In our example this might mean finding out what other diseases the patient has, extending the range of symptoms in accordance with the needs of a finer diagnosis, etc.

The schema of 'situations', as follows from this account, can be converted into the schema of 'thing – property – relation'. The same can also be seen from the relation between the two methods which are associated with the schema of 'situations', namely those of structure-description and state-description.[3] In structure-description we assume that a given universe is decomposed into a (finite) number of classes. The core of structure-description is the *establishment of the number* of individual objects, pairs, triplets,... *n*-ples of objects belonging to the respective classes. State-description assumes in addition that we are capable of distinguishing each individual object. Hence the core of state-description lies in the fact that we determine which *particular* individual objects, pairs, triplets,... n-ples of objects belong to the various classes. By presenting the structure description and also the state description of a given universe, we in fact furnish a certain picture of the situation of that universe.

(c) In either of these two schemata of the internal analysis of a universe, we generally assume that we are in a position to scan, or specify clearly, the entire universe. In contrast to this, there is yet another schema we encounter frequently in both day-to-day life and science, where this prerequisite is not satisfied. If we say 'it is raining here now', we are not giving the meteorological situation for the whole of our geographical area, but are merely making a statement about a certain event, at a certain time and in a certain place. This mode of expression has a schema which we may call the *schema of 'events'*.

The schema of 'events' is frequently used in the expression of the results of observation, measuring or experiment, especially if the expression has the nature of a *Protokollsatz* or a kind of *Konstatierung*. The elementary form of a *Protokollsatz* might be illustrated by the sentence: 'The observer (experimenter) found this, that and the other at a given place and given time'. It follows that this says nothing at all about events at other places or times nor about anything that the observer might have found in the same place previously or at the same time elsewhere. Such discoveries would call for further *Protokollsätze*.

These various schemata do not of course represent all the schemata possible which we encounter in everyday speech and in the language of science. It does seem, however, that they represent the most frequent and most important ways of handling the things we consider as the objects of cognition in the empirical and experimental sciences. The results of cognition can be expressed in various schemata, and it is also possible that one and the same result can be expressed using different schemata, or that one schema can be reduced to another. We have already given a partial indication that it is relatively easy to convert the schemata of 'situations' and 'events' to those of 'thing – property' or 'thing – property – relation' respectively. For this reason the following account will be based on the schema 'thing – property – relation'. This choice also governs the adaptation of the language devices used.

All of the schemata of analysis of a universe which we have been mentioning here, and probably others not here considered, are tied up with difficult problems of a semantic and ontological nature which are bones of contention between nominalism, Platonism and conceptualism. Since it is not the aim of this work to analyze these questions, we shall leave them aside. It should, however, be borne in mind that the choice of the schema selected is contingent and relative.

2. The communication model and the problem of scientific empiricism

(a) *The Communication Model of Cognitive Activity*

The results of cognitive activity in science having an informatory character, they are only communicable by means of language, as statements made in language. Statements expressing these results are always of a *mediated character*. The mediation lies in the fact that the objects of a universe – whether these be objects distinguishable by the senses or any other objects to which we relate statements – are accessible only on the basis of certain procedures.

For the sake of illustrating this mediation, we may use to advantage the communication model, which assumes a communicatory linking of at least three blocks: block (A) 'the source of information', (B) that of the 'channel of the observer', 'the channel of the experimenter or measuring apparatus', or even more generally the 'channel of science', (C) the block of 'statements'. Diagrammatically the combination of the three blocks might be represented as in Figure 1.

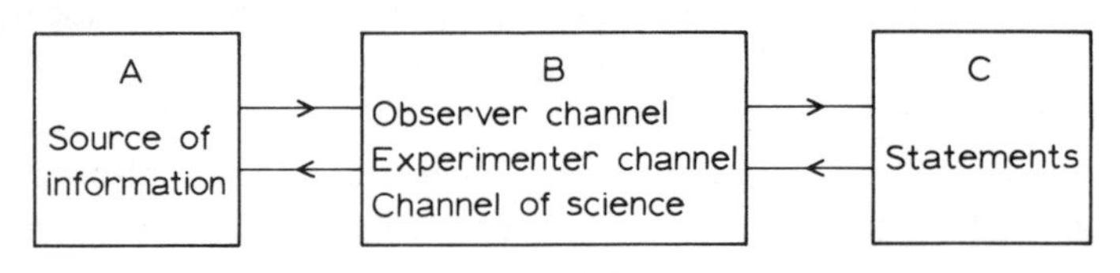

Fig. 1.

Block A may be pictured as a (finite or infinite) set of objects or events of the most various kinds. Simultaneously it is assumed that at least some of these objects are distinguishable by block B. In other words, block A comprises a set of objects which can be taken as distinguishable stimuli for block B. They might, for instance, be a set of objects distinguishable by the senses and observable, where block B is a human observer. Or we might conceive of a set of objects which can be reconstructed as the (hypothetical) causes behind certain reactions of block B, such as a set of ailments which cause certain distinct symptoms, or the movements of microparticles which leave particular distinct marks in the measuring or experimental apparatus, and so forth.

Block B may be a human observer, or one furnished with apparatus which extends the discriminatory capacity of the senses (e.g. microscope, telescope, measuring apparatus), or any measuring or experimental apparatus capable of recording the stimuli emanating from block A. By saying that block B is capable of distinguishing the stimuli emanating from A, we are not saying that it is inevitably capable of distinguishing them all, nor that it necessarily distinguishes them adequately. We assume that there is a communication link between blocks A and B, but also that it may be marred by noise, which can have an effect on the degree of precision and the error-rate which characterize the reactions of B to stimuli from A.

There are certain important differences between the properties of block A and those of block B. While block A may be an infinite set of objects or events and is hence capable of generating an infinite amount of stimuli, block B, in a certain time and space, is never capable of recording more than a finite number of stimuli. The problems associated with methodological finitism are thus seen to be rooted in block B and not necessarily in block A. Moreover, block B is always characterized by the *thresholds* between which its ability to distinguish stimuli coming from A is operative, as well as by a finite and restricted delicacy in its power to distinguish. This means that the activity of B always has certain *limitations*.

These, however, are not the only limitations. Others may be assumed as being connected with the possibility of (uncontrollable) retroaction of block B on block A. (The possibility of such retroaction is indicated in the diagram by the two-way linking of A and B.) From the evolution of scientific cognition we know that limitations of this kind were encountered in quantum theory, where they were formulated by way of the relation of uncertainty: if the area of a micro-particle can be measured with an accuracy of Δx, and if its impulse can be measured simultaneously with an accuracy of Δp, then the familiar relation

$$\Delta x \times \Delta p \geqslant k$$

holds, where k is a quantity of the same order as Planck's constant h. Analogous relations have also been formulated for other pairs of physical quantities. It is obvious that the analysis of the state of block A by means of block B is possible only within certain limitations, the latter being due to the retroaction of B on A. (An account of the uncertainty relation might take as block B the electron microscope. In that case, the measurement of the time-space and impulse-energy parameters of the object measured must be affected by the properties of the rays used in the electron microscope, as we approach a certain limit.)

Another form of such limitations, analogous to Heisenberg's uncertainty relation, has been pointed out by Gabor (Gabor, 1946; p. 429) in the analysis of acoustic measuring. If the signals are measured using apparatus with a bandwidth of ΔF, then the shortest measurable signal is ΔT, the product of ΔF and ΔT again being equal to or greater than the constant of the order unity.

The feedback between B and A may, naturally, be of very different characters. For example, it may take the form of power action of the measuring apparatus on the object being measured, causing changes in the states of that object in such a way that at the output of block B we do not receive information which is reliable for estimating the state of A, but information on the modified state of A as distinct from the original state before the feedback took effect.

The following consideration leads us to the conviction that this analysis of mediation as based on the communication model may be of significance quite generally and need not be confined merely to measuring in the natural sciences and technology: Let us assume we are to organize a sociological survey of attitudes, public opinion, a prestige scale, etc. When formulating the questionnaire, we may present some of the questions so suggestively that the attitudes of the persons questioned may be somewhat influenced. It is sufficient to use certain terms with an emotional content, or to include questions which in themselves affect the kind of answer. Hence we see that in this case, too, block B alters the original state of block A, rendering the data obtained at the output of B inadequate for A, or making them the result of the interactions of blocks A and B.

We assume that blocks A and B have their own structure: this means, for example, that the measuring apparatus which we use to obtain information about an object under investigation need not be of the same nature as that object; it may have its own independent structure. At the same time we assume of A, quite naturally, that it exists, or is established for the purposes of the relevant investigation, as objective, i.e. as independent of B. On the other hand, it is equally certain that any statements about anything affecting A are possible exclusively on the basis of the data available at the output of block B. This also means that scientific empiricism cannot accept any considerations of the kind based on mediated 'getting to the heart of the matter', 'vision of the essence', 'intuitive comprehension of the essence', and suchlike.

It is likewise true of block C that it has a relatively independent structure: it is a system of statements based, on the one hand, on data at the output of block B, and on the other hand, on different statements included in the system but taken from other sources. We are essentially dealing here with the two kinds of mediation pointed out by Russell, who distinguishes between knowledge by acquaintance and knowledge on the basis of description.[4] Knowledge by acquaintance is here understood as secondary knowledge, mediated by block B, irrespective of whether this block contains merely a human observer having normal sensory equipment or whether he is furnished in addition with measuring or experimental apparatus. Knowledge on the basis of description must, naturally, be also based on observation, measuring, experiments, etc., but anyone having block C at his disposal comes by these data (descriptions, in Russell's terminology) by virtue of processes of communication with other people, scientists, experimenters, etc.

This dual mediation might be illustrated diagramatically by the crossing of two communication processes (see Figure 2).

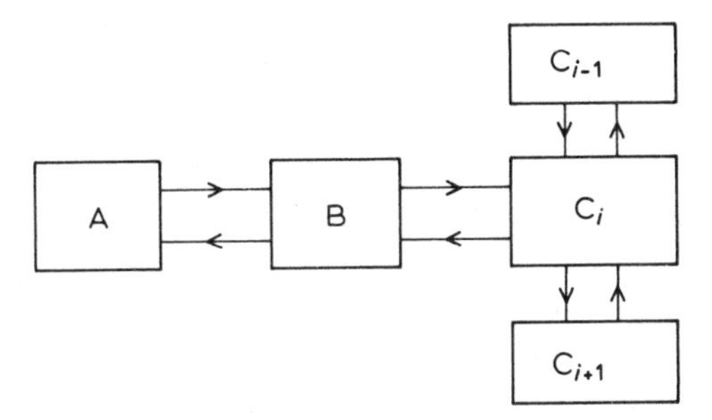

Fig. 2.

The relations between C_{i-1}, C_i, C_{i+1}, etc. may be interpreted as the everyday processes of communication between two or more communicants. This may take the form of a dialogue, or the relation between the author and a reader of a scientific study, or that between a lecturer and a member of his audience, and so on.

When we say that block C (and the same will apply to all other like blocks, i.e. C_{i-1}, C_i, C_{i+1}, etc.) relates to statements, it means that when we come to formulate them we must respect the respective syntactic, semantic and other rules which are binding for the language in which the statements are made. Here too, emphasis on the relatively independent structure of C remains valid.

The differentiation of blocks A, B and C of the communication model of cognitive activity further makes it possible to give more precise expression to the distinction of ontological and semantic decision. Where decision-making concerns entities distinguishable in block A, we may speak of ontological decision. This means that ontological decision relates (accepting the schema of analysis 'thing – property – relation') to facts and the things, properties and relations which are distinguishable in them. Semantic decision relates to the statements and their components which we consider as elements of block C. Methodological decision-making, which stands on rather a different level, relates to means and procedures, their selection as adequate, their optimization, and so forth. This means that methodological decision-making is connected with the activity of block B, not only with respect to blocks A and C, but also to the goals of the given task and to what is expected from its solution. In other words, methodological decision-making cannot but respect ontological and semantic questions, but it must at the same time involve pragmatic viewpoints.

(b) *The Communication Model and the Problem of Semantic Analysis*

The communication model here described, which emphasizes two kinds of mediation, makes possible a new and more fruitful interpretation of some of the traditional concepts with which we work in the analysis of the statements of science. Those which we have particularly in mind are 'assignment' (in the semantic sense), and 'level' and 'experience'.

Let us look first at the concept of 'assignment'. This is usually encountered in simple and, up to a point, naive questions of the kind

'What corresponds to ...?', 'To what does ... relate?', 'What is denoted by ...?' and suchlike, where '...' are statements which we consider as elements of block C. These and similar questions require in effect that elements of block C be assigned in a simple way to elements of block A. The naivety of questions framed in this way lies in the requirement that elements of block A be assignable to each of these statements and their component parts. To put it another way, it is required that all the components of the statements have denotation represented by objects which constitute elements of block A. This requirement conflicts, of course, with the relatively independent structure of C that we have been emphasizing.

Since each of the statements that are elements of block C is composed of signs (or verbal signs, words), we may, from the point of view of the relations between blocks A and C, draw the distinction between proper signs (or categorematic words, or extralogical words, as the case may be) to which elements of block A are assignable, and improper signs (or syncategorematic, or, in Russell's terminology logical words, as the case may be) to which elements of block A are not assignable. The requirement that all signs be assignable thus means that we ignore the relatively independent structure of C and that we fail to allow for the fact that any statement that may be an element of C is only possible in the framework of a particular language and that that language always has a syntactic structure.

Leaving aside improper signs, whose role is connected with the syntactic structure of the language used, the analysis of statements and their components, which operates with elements of block A, finds its place among the traditional problems of denotation. However, these problems, which involve the assignment of elements of block A and block B, cannot be solved in each case by a one-to-one assignment of the kind called for by questions in the naive form of 'What corresponds to ...?', 'What does ... refer to?', and suchlike. Cases do, for example, occur where what are assigned to elements of block C are not individual elements of block A but certain relations or dependences of these elements. Moreover, due respect must be paid to the fact that the mediating role of block B often takes effect in the assignment.

It goes without saying that the semantic analysis of any statements and their components which we take as elements of block C cannot be merely reduced to denotation. Not only denotation but also the sense of the statements and their components is important here, which is of significance especially in connection with the processes of communication between C_{i-1}, C_i, C_{i+1}, etc., i.e. what corresponds to Russell's conception of knowledge by description. The fact that statements have sense is what enables us to understand them. From the point of view of the processes of communication between C_{i-1}, C_i, C_{i+1}, etc., the sense is what is preserved in the case of a reliable communication established between any of them.[5]

A major problem to which attention must be paid in any semantic analysis of scientific statements is that of 'level' and the selection of the appropriate level for assignment. If we assume that block A is made up of certain elements, we must always allow for the possibility that they may be in some way categorized, ordered in a hierarchy, and so on. What we have described as various schemata of analysis of a universe is also relevant here. Using, for example, the schema 'thing – property – relation', we may, though it does represent a kind of reconstruction *ex post*, count on a conception of assignment whereby 'things' are assigned to individual expressions (e.g. individual descriptions, the proper names of individual objects), 'properties' and 'relations' to predicates, and so forth. Here, as in the case of other schemata, the question arises – and it is entirely justified – as to whether we are not, after all, assigning things which are proper to block C to block A as well. In other words, fears may present themselves that we are imputing to the very universe a structure which is only proper to language and statements in language. With any reconstruction of block A, and generally with any reconstruction of the universe, due account must be taken of the fact that there is feedback between blocks A, B and C, and that it may also figure as the direct influence of block C on block A. This means that in any such reconstruction we must proceed very critically, checking each step and confronting the results of each step against each other. It is of particular importance to guard against naively realist illusions of the type which usually betray themselves in the effort to assign 'something' in block A to every element of block C.

The emphasis placed on the fact that block C has a relatively independent structure may also be interpreted as meaning that this structure is one of an ordered hierarchy. Having said that block C is made up of statements, we must also take account of the fact that for their formulation both proper and improper signs of the given language must be used. Also, these statements, which must be formulated in strict accordance with the relevant rules of the given language, may go to form certain wholes. And so on. This brings us to a question which might be formulated as follows: Which of the categories on which the formulation of the statements, or larger wholes made up of statements, are based may justifiably be considered as that where it is appropriate to speak of assignment? We have indicated that we tend as a rule to assign things to individual expressions, and certain properties and relations of these things to predicates. Are we, however, justified in considering in a like manner assignment in other, especially 'higher', categories? The traditional logico-semantic solution to the problem effectively considers only the relations between blocks A and C. Yet if we consider its truth value, i.e. an entity of a constructive nature, as the denotation of a sentence, then we are already beyond the sphere of the relations between blocks A and C. (The same applies to other attempts at solutions of sentence denotation, namely those operating with concepts such as 'belief', 'assertion', 'conviction', etc.) In principle, of course, there need be no objection to the conception of truth value as sentence denotation. What must be borne in mind, however, is that this conception cannot be placed on a level with that of the denotation of individuals and predicates, and also that it exceeds the sphere of the relations between blocks A and C. In practice this means that we are actually operating in this sphere at only one of the levels, while in the cases of the other levels, i.e. those of statements or sentences, sentence-wholes, etc., we are already outside it. Naturally, if we select a different schema of analysis of the universe, there will also be a change in the form of the level which remains within the sphere of the relations between blocks A and C.

(c) *The Problem of Levels*

So far we have been considering the concept of 'level' from the point of view of block C, the sphere of statements. It can, of course, be equally well considered from that of block A. In this respect and according to the kind of phenomena which are to be considered, we generally find a differentiation of levels: the macro-level and micro-level, the macroscopic, quantum and sub-quantum levels in physics and chemistry, the factory, industry and national levels in economics, and the various levels familiar from biology, the social sciences or any others. In every one of these cases the concept of 'level' relates to block A or to the respective universe generally. It is evident, furthermore, that the distinction of the various levels is based on objective differences and more or less objective criteria.

By emphasizing that levels are distinguished on the basis of objective differences (such as exist, for instance, between quantum processes and macro-processes in physics, and also, quite naturally, between the descriptions of the two kinds of processes), we do not wish to say that the concept of 'level' does or can have an absolute significance, nor that it is possible to reconstruct a uniform scale of levels which will be absolutely valid for every situation. There have been many attempts to set up such an absolute scale, especially in the science of the 19th century, but these usually lay behind the endeavour to find an objective reason for the classification of the sciences.

The view that the distinction of levels is not absolute also means that it is not possible to demonstrate the existence of some fundamental, elementary or initial level as a basis of reduction for the others, i.e. a level to which all formations of the higher levels can be reduced without remainder. This is not, of course, to deny the significance of any such reduction, still less to deny the explanatory value of it. To take the example of the set of molecular, atomic and quantum levels: it is a familiar fact that modern physics explains many phenomena which appear at a 'higher' level by elements and means demonstrable at a 'lower' level. Analogous situations are to be found in the relations between the levels used in biology, biochemistry, chemisty and other branches of science. However, it is another familiar fact that certain phenomena, properties or processes which appear at a higher level cannot be explained in their entirety by all that is known or available to us at the 'lower' level. Two reasons for this may be, on the one hand, our insufficient knowledge of the 'lower' level, and on the other hand, the occurrence of certain new, qualitatively distinct phenomena at the higher level.[6] In other words, in addition to quantitative differences between the various levels, allowance must also be made for possible qualitative differences.

It follows from what we have said about the relations of the various levels that we assume a multiplicity of scales of levels, and that it is always relative what, under given circumstances, is taken as the base level. When a relative base level is to be determined, allowance must also be made for the possibly major role played by block B, the channel of the observer, measuring apparatus, or generally the channel of science. Any, even the most perfect, system where there is a part played in the cognitive process by block B is restricted in the degree to which the objects in block A can be distinguished, i.e. it has certain thresholds, a finite fineness in its power to distinguish, and so on. Thus the concept of 'level', in the sense in which we have just been considering it, is always relativized to the properties and potential of block B. So from this point of view, each level which we are capable of distinguishing from another, irrespective of whether 'higher' or 'lower', is associated with no more than the ability to distinguish a finite number of classes or a finite number of properties assigned to the objects of block A. In this lie the very foundations of methodological finitism.

It remains to explain what is meant by 'higher' and 'lower' levels. This distinction, too, must be relativized to the properties and potential of block B, since it depends on the range of the data coming from block A and which block B is capable of distinguishing. If, for instance, block B is capable of distinguishing data $a_1, a_2, a_3, \dots a_n$, and if we then improve the discriminatory capacity of block B (which in fact means a shift from block B to block B′) so that in the domain (e.g. of time or space) where previously a_1 and a_2 were distinguishable we can now distinguish $a_{11}, a_{12}, a_{13}, \dots a_{1n}$, and $a_{21}, a_{22}, a_{23}, \dots a_{2n}$, then we have passed from a 'higher' to a 'lower' level. It is natural that the elements of the higher level do not always coincide with the sets of elements of the lower level.

From the point of view outlined here, any refinement of the discriminatory ability of block B is a transition from a higher to a lower level, and the reverse, which is usually accompanied by a broadening of the field of vision, is a transition from a lower to a higher level. However, in the thinking of the natural sciences the conception of levels has become stabilized in a way that conditions for the transition from one level to another are not seen in merely any change in the properties or potential of block B, but only in such changes as lead to a qualitative change in the nature and function of that block.

There are three main facts characteristic of a transition from one level to another.

(a) In any case of such a transition, we usually come up against the limits of the potential of apparatus of one type. The case of the limits of the potential of macroscopic measuring instruments with respect to micro-processes is expressed by the Heisenberg relation, to take an example.

(b) With any such transition from one level to another there is a change in the character of the fundamental regularities of the phenomena investigated. To remain with the previous problem, it is obvious that the nature of the regularities of quantum physics is different from those of classical mechanics.

(c) A transition from one level to another is also accompanied by a change of the unit of measurement or of the set of fundamental parameters for characterizing the phenomena investigated.

Although it seems that these facts can be best demonstrated on the example of transitions from level to level in physics, analogous facts probably also present themselves in other domains. However, they are not always respected to the same degree as in physics. It is, for example, quite essential in the analysis of economic problems to have available distinct sets of basic parameters according as the point at issue is the economic behaviour of an individual, economic problems within a single enterprise, or economic questions in the framework of the state economy.

(d) *The Communication Model and the Concept of 'Experience'*

The communication model of cognitive activity which we have been describing also enables us to give a more accurate interpretation of the concept of 'experience'. In the traditional view, experience was understood to mean the sum total of sensory data, i.e. data which the human observer is capable of distinguishing and registering. Taking the communication model as a basis, with its blocks A, B and C, the following basic inadequacies of the traditional view come to light:

(a) The traditional view of experience is generally very static. It confines experience to the aggregate of ready data. The very term 'data' itself, which is nowadays generally understood in the informational sense, originally used to express something set, 'given'.

Modern science, however, cannot afford to confine experience merely to a certain ready-made and always complete set of data qualified in some particular way. Experience, as we undertstand it intuitively in modern science, is more a process, a sequence of operations, deliberate steps and so on.

(b) Another weakness of the traditional conception of experience is that it deliberately abides by data which would by characterizable in the same way for all situations. This means in effect that the traditional view works with only one type of block B binding for all situations. This leads to the attempts at a uniform characterization of experience. It is, of course, a familiar fact that such attempts, as offered, for example, by pragmatism, operationalism and some versions of logical empiricism (e.g. the conception of what were called *Protokollsätze*, '*Basissätze*', '*Konstatierung*', and so forth) could not be considered satisfactory for all situations.

(c) The traditional conception of experience identifies experience with the aggregate of ready data, which means in effect that any data at the output of block B are taken. This in turn means that no consideration is given to the varying degrees of relevance of these data. The concept of the relevance of data is not, of course, intended abstractly; it is always bound to a given task or class of tasks and to the goals which are connected with their solution. This means, in fact, that not all of the recordable data are essential to the solution of any given task. In diagnosis, the doctor may obtain a great many data about the patient, but for the immediate purposes only certain symptoms are of importance, while here too the degree of relevance of the symptoms need not be the same. Hence the concept of the relevance of data cannot be understood merely semantically as bound to a given universe, but also pragmatically as bound to the goals, requirements or expectations which we connect with the solution of the relevant task.

Cognition in science, especially in the natural sciences, operates in practice with a different, much broader concept of 'experience'.

The traditional account of the conception of experience has been developed mainly in philosophy. The modern conception is rather to be found in the theory of statistical decisions or in game theory. Assuming that we are operating with the by now familiar blocks A, B and C, then the last of them involves a *decision process* the object of which is to estimate the state of block A. In general terms, experience is understood to mean all steps leading up to the refinement of the estimation and the improvement of the decision-making. It is obvious that this conception assumes that experience is not a static act but a dynamic process, a sequence of steps.

This conception of experience can be readily demonstrated on a simple example of statistical decision-making. Let A be an urn containing a large number of black and white balls. We cannot see inside the urn, and we can only take a small number of balls out of the urn at a time. Clearly, each additional handful goes to improving our estimation of the original *a priori* distribution, or to giving greater accuracy to our hypotheses on the source of information. A system whereby we can calculate with the results of previous steps (i.e. previous handfuls) might be described as an experience system. This kind of system assumes that further steps are taken on the basis of the evaluation of the previous steps. This conception clearly does not take just any data in isolation and irrespective of the tasks in which the data may be used, nor irrespective of the goals which the solution is intended to attain. Experiential data in the traditional view thus represent experience only in so far as they go towards improving our estimations, refining the hypotheses on the basis of which decisions are made, and so to raising the standard of the decision-making.

The example here given of an experiential system can be given further modification. As a rule, we do not study in the empirical sciences only those domains of objects which remain invariable, but also processes of change. From this point of view, it is then possible to alter the original examples of a source of information slightly. Let block A be once more represented by an urn containing a large number of black and white balls. We are to estimate the ratio between them. Again we assume that we cannot see inside the urn and that only a small number of balls can be taken out at a time. Let us now imagine that not only a human observer or experimenter is manipulating the urn, but also a demon, using for example a different hole into the urn. This demon is also able to add or extract the balls from the urn, likewise step by step and in limited quantities, and so change the original distribution. One of two situations may then arise: Either the demon 'works' more quickly than the human observer, who thus does not have the chance to improve his estimation of the original ratio. This sceptical view of the experience problem is adopted in Bergson's conception of scientific cognition. Developments in the natural sciences, however, tend rather to confirm the second case, in which it is the human observer who works at the quicker rate and is more prompt and also more cunning than any demon of nature. In this view, the experiential process is a game with a partner whose strategy is unknown, or only partially known, and who is in a position to change his strategy up to a point, but whom we are in turn capable of overcoming each time. This latter case corresponds to the conception of scientific cognition coined by Einstein and Norbert Wiener.

From these two sketchy outlines of experiential systems it follows that experience is not and cannot be merely data at the output of block B, i.e. the unevaluated results of observation, measurement and experiments, but also the evaluation of previous results of this kind as well as of previous decisions made on the basis of earlier results. Hence, too, the concept of experience includes not only data at the output of block B but also any measures taken leading up to an improvement in the standard of decision-making, such as the improvement of the functioning of block B, the improvement of the hypotheses on the basis of which decisions are made, closer estimations of the state of or changes in block A, as well as the clearer expression of the expectation of possible consequences from the decisions selected. Thus for the broadening and improvement of experience not only progress in the sense of the quantitative growth of empirical data is of essential importance, but also any development of a theoretical nature, or developments in the techniques of measurement and experiment capable of leading to improved decision-making and a greater degree of accuracy in our knowledge of the universe we are investigating.

3. The Universe and the Language Used

(a) *The Entities of the Universe*

An essential prerequisite to the specification of the language of science to be used is the fixing of the relevant universe. Examples of such a universe might be all the heavenly bodies (irrespective of whether or not they have already been distinguished, observed and described using the means which we have at our disposal to date), all living creatures, all electromagnetic phenomena, and so forth. For the purposes of our further analysis of scientific procedures, our basis will be the schema 'thing – property – relation'. Hence we shall be assuming that the objects of a given universe which we shall be considering as 'things' are distinguishable by virtue of the fact that certain properties and relations pertain to them, the circumstance that certain properties and relations do pertain to them being called a fact.

Hence, we shall consider facts, things (objects), properties and relations as the basic entities of a universe. The construction of the vocabulary of the language used, in particular the proper signs (verbal signs or words), must also be adapted to this choice. It goes without saying that if we were to select a different schema of analysis of the universe, the construction of the vocabulary would have to be adapted accordingly.

In addition to the basic entities, we may also allow of what might be called derived entities, which we construct on the basis of the basic entities, such as classes of things, classes of properties, or classes of relations. We mention the construction of derived entities for the reason that it invariably presupposes a certain level of cognitive, and above all abstractive activity. In terms of the communication model, the elements of block A will be what we have described as the basic entities. This also means that the problems of assignment (in the semantic sense) in the case of derived entities cannot be solved directly but inevitably via the basic entities.

(b) *The Language Devices Used*

The language devices which we shall be using are based on the following vocabulary, the proper signs of which are:

(1) Individuals, which we will assign to 'things' and denote by lower case letters, e.g. a, b, c, together with numerical indices as required. We shall use x, y, u, v, and z as individual variables.

(2) Predicates, which we shall assign (according to the number of arguments) to properties and relations and denote by upper case letters, e.g. A, B, P, Q, R, using numerical or alphabetical indices as required.

(3) Statements, which we shall assign to facts and denote by italic capitals, e.g. S, E, T, with numerical or alphabetical indices as required.

As improper signs we shall be using: $\sim$ (negation), . (conjunction), $\vee$ (disjunction), $\equiv$ (equivalence), $\rightarrow$ (material implication), / (incompatibility), $(\forall x)$ (general quantifier), $(\exists x)$ (existential quantifier), and $\in$ (sign for membership in a class).

To denote classes we shall be using ronde capitals, e.g. $\mathscr{X}$, $\mathscr{A}$, $\mathscr{P}$. When assuming the construction of a class on the basis of P, i.e. assuming the class of all things having the property denoted as P, we shall use the sign $(\hat{x})\, Px$.

The selection of syntactic means corresponding to the rules of elementary predicate logic must also be adapted to the given conception of a universe and its entities, as well as to the vocabulary. Since by a fact we understand the circumstance that certain properties and relations pertain to things (some, no, all things), we shall consider as statements all closed formulae which concatenate predicates with individuals according to the relevant number of arguments.

Any further signs which will appear here and there throughout the account will be introduced in due form in the appropriate context.

NOTES

[1] It was Körner who indicated the possibility of using various schemata for the analysis of a universe (Körner, 1966).

[2] Thinking consistently, we find that the concept of 'thing' is an abstraction in a similar respect as is the concept of 'property'. The fact of the matter is that in the continuum of distinguishable phenomena we tend to consider some as 'things' and others as 'properties'. Nor is it out of the question that what we call a 'thing' in one connection will be called a 'property' in another.

[3] The fundamental ideas of the method of state description were laid down by Wittgenstein (1922). The methods of state-description (as the background to semantic analysis) and structure-description have been elaborated by, in particular, R. Carnap in his works on logical semantics.

[4] See in particular Russell (1905, 1919, 1948) and certain others of his works.

[5] The concept of 'sense of an expression' may also be interpreted in a variety of ways, for example, as a constructive category, as the relations between statements and their components on the one hand and these categories on the other, as an invariant which remains preserved in the case of communication, translation, etc. It is likewise possible to construct various criteria for sense, e.g. empirical criteria, translational criteria, and others. For a more detailed analysis of these problems see Tondl (1966b).

[6] Adherents to emergent philosophy speak in this connection of the 'emergence' of new properties and processes and hence of 'emergent' properties. However, they often forget that these 'emergent properties' may be merely due to our inadequate knowledge of the 'lower' levels, or, as some conceptions of physics put it, to the existence of 'hidden' parameters. This does not, of course, mean that hidden parameters must be postulated at all costs and in all circumstances as has been attempted by one school of interpretation of the phenomena of quantum mechanics.

CHAPTER IV

CONCEPTS OF THE LANGUAGE OF SCIENCE

1. Names, descriptions and statements

(a) *Singular and Divided Reference*

In the language of science the distinction is drawn between those verbal signs which may be assigned to individual things (individual objects) and those which may be assigned to classes. The former include, for example, 'Mont Blanc', but also 'the highest mountain in Europe', 'Karel Čapek', but also "the author of the play 'RUR'", etc. The second group includes, for example, 'water', 'atom', 'metal', 'nation', etc. It is obvious that any more precise differentiation of the two kinds of signs is both difficult and problematic. Quine [1] distinguishes in a similar sense between singular and general terms. A general term, according to Quine, is one which allows of being used with the definite or indefinite article and in the plural. He concedes, however, that from the semantic point of view this distinction is naturally vague. Another of the characteristic features of general terms lies, again according to Quine, in the fact that they have what is called divided reference. The term 'gas' may denote this particular gas, some gas or other, any gas, all possible gases, and so on. On the other hand, however, the expression 'the first man to fly in space' can denote but a single person, and that person was Yuri Gagarin.

It is natural that any considerations as to singular and divided reference are possible on the assumption that the universe is fixed and that we are capable of distinguishing the various objects in the universe such that it is possible to distinguish between objects and classes of objects. It also requires the assumption that we respect the extensional approach to verbal signs. If these assumptions are not made, Quine's distinction between singular and general terms ceases to have sense. In particular it is dubious whether it is possible to apply the extensional approach conceived in the same way for all expressions.

Quine holds that the difference in the roles of singular and general terms is to be seen in what he calls the process of predication. If we say that *a* is P (or P*a*), then in the simplest case we assume that '*a*' is a singular and 'P' a general term. It is of course evident that such an assumption concerns only the typical mode of predication which is but one of the elementary modes of predication. Moreover, as Quine himself notes, in neutral logical notation the 'P' of the predication 'P*a*' may be represented by a noun, adjective (in either case the copula 'is' is essential) or verb; each of these types may be considered as the verbal form of the predicate.

(b) *Proper Names and Descriptions*

The distinction as described above between verbal signs with singular and those with divided reference should not be confused with another differentiation: it is necessary to distinguish between verbal signs having the nature of proper names and those which may be characterized as descriptions. It will have been seen from the examples already given that one and the same geographical object may be denoted by the proper name 'Mont Blanc' or by the words 'the highest mountain in Europe'. It was on this differentiation that Bertrand Russell founded his theory of descriptions. While proper names are simple verbal signs which denote individual objects directly, descriptions may consist of more than one verbal signs, their meaning being given by the actual way in which the signs used are combined. Like the foregoing this differentiation is also considerably vague and in many cases not very reliable.[2] For if we are to accept Russell's differentiation of names and descriptions, we cannot overlook the fact that a considerable proportion of names – considered etymologically – were originally descriptions or arose out of descriptions. For this reason any more precise differentiation of names and descriptions is also more a matter of an appropriate convention.

The differentiation of names and descriptions may become important if we accept the view that the name-relation relates only to names while not affecting descriptions at all, or, to put it another way, that it is possible to talk in terms of denotation only in the case of names, while in that of descriptions we may speak of sense and in some cases also of denotation. This view has been formulated best by Ryle.[3] His main protest is against the name-relation applicable in the case of names being applied in the semantic analysis of descriptions. Names, more precisely proper names, are the naming of persons or objects and not descriptions of persons or objects: Winston Churchill was the father of Randolph Churchill. He was, however, named and christened 'Winston Churchill' and not 'the father of Randolph Churchill'. On the other hand conceptions which treat descriptions as if they were names – Ryle in particular mentions those of Frege, Moore and Russell, but his criticism also affects the dualistic semantic conception represented by Carnap, Church and others – in fact identify the concepts 'to be a name' and 'to have a sense'. The confusion of these two concepts, which it is quite essential to distinguish, is expressed by Ryle in the principle 'Fido' – Fido: if 'Fido' is the name of a particular dog then 'Fido' denotes the dog Fido. If we then assume that the word 'dog' denotes objects in the same way that 'Fido' denotes Fido, it must denote the class of all real and imaginable dogs, or the class of all canine properties which may be shared by all dogs. This is the way, as Ryle shows, in which all *denotata* or *nominata* of words are constructed, words, that is, which are not names but descriptions. This gives rise to the denotational theory of sense which is then obliged to assign to each word or complex of words having the character of a description (and these need not be merely nouns but also adjectives or verbs) either an assumed or merely hypostatized denotatum, i.e. object or class of objects 'denoted' by the given expression. The expression 'the first man to set foot on the Moon' concerns an actual person, namely the American citizen Neil Armstrong. However, the expression 'the first man to set foot on the Moon' is not this citizen's name. It merely describes one of his properties. This description may be related to only one object. Other descriptions may be related to a number of objects, to a certain class of objects, etc. It is quite natural then that such expressions in language which as descriptions cannot be related to any object are also perfectly justified. Thus, for example, 'the first man to set foot on Mars' is an expression which we can readily understand, having for us a quite definite sense. However, if we were to proceed on the 'Fido' – Fido principle, we should have to conclude that this expression cannot be related to any object. (Adherents of the denotational conception, who assign to every description a class of objects, are then obliged to state that while the given expression is indeed the name of a class, this class is empty.)

Ryle's criticism of the denotational conception and of the hypostatization of the special entities which are the denotates of words which are not names, is an expression of extreme nominalism in the conception of the language of science. According to this viewpoint proper names and proper names alone may denote objects. Expressions such as 'dog', 'atom', 'red', 'fulfil' 'be greater than', etc., i.e. expressions having the character of predicates, do have a sense but do not have denotation, that is they do not denote anything. Therefore from the point of view of extreme nominalism these expressions have the character of syncategorematic words.

There are certain objections which may be drawn against this viewpoint, and these are backed up by the actual use of these words in everyday speech and in the language of science:

(1) In everyday speech these words are usually used as elements of what Russell calls denoting phrases: If I say that a dog has bitten me, it is usually evident from the context of the conversation that I am speaking of a particular dog, for instance that of our neighbours. The denotatum of such expressions may be more precisely specified by stringing together the given word or complex of words with other words, for example 'our dog', 'the black dog I saw in the street', 'all the dogs in our town', etc. In none of these or similar cases is there any absolute need to hypostatize a special entity such as the class of all dogs present, past and future, the class of all canine properties, and so on, for the simple reason that such an entity is not what is the subject of the conversation, which, as a rule, concerns only certain objects. The range of objects meant is more or less determined by the context of the conversation or by the use of additional specifying expressions.

(2) Likewise in the language of science these and similar words or word complexes are used in such a way that as a rule it is evident what these words denote, i.e. what are their denotata. The decision as to what is actually denoted by them usually depends on certain assumptions which in the language of science are either explicitly or sometimes only tacitly respected:

(a) The universe to which the statements refer is always fixed. This universe need not always be a set of empirically distinguishable objects. It may also be a class of mathematical objects, a system of numbers, etc. The acceptance of a certain universe may then be justified in a variety of ways, for example on the basis of accepted empirical criteria, of criteria of a theoretical nature, or of certain appropriate conventions.[4] Thus when deciding what given words or word complexes denote it is necessary to respect the universe accepted as fixed.[5]

(b) Another assumption respected in the language of science is that of the possibilities of decomposition of the universe. This means that we are capable of distinguishing individual objects and classes of objects, ordering the individual objects and classes in a particular way, etc. It also means that we are capable of speaking of all, some, one or none of the objects of a given class and that we are capable of operating with variables. In particular it was Quine who pointed out the importance of variables for fixing a certain universe and for the possibilities of its decomposition.[6]

It is evident then that in the language of science the recognition of the fact that a given word or complex of words denotes something, i.e. that it is assignable to a certain object which is the denotatum of that word, need not necessarily mean that this object is an empirically distinguishable individual which may be the bearer of a proper name. Thus although fears that the name-relation may be extended without justification to words having the character of descriptions are up to a point understandable and justified in the same way as the criticism of the platonizing hypostatization of abstract entities is justified, this need not mean that the semantic analysis of words having the character of descriptions – analysis on the basis of the name-relation – is in any way prohibited.

If it is possible, in the case of words having the function of descriptions, to apply semantic analysis in such a way that sense and denotation can be assigned to the words, then the situation is quite different in the case of words having the nature of proper names. That this latter situation is different becomes evident when we explore the possibilities of translation: Descriptions can and usually must be translated. The expression 'the first man to set foot on the Moon' can be translated into all natural languages. On the other hand, however, the name of the first astronaut to set foot on the Moon's surface, i.e. the name 'Neil Armstrong', is never translated as a matter of principle.[7] If we take the sense of words and word complexes to be what is retained in any correct translation of the words, then it will be seen that in the case of words which are beyond all doubt of the nature of proper names it is relatively difficult to talk of the sense of these expressions.[8]

If the results of cognitive activity are expressed in natural language, it should be borne in mind that the differences between words having the character of proper names and those having the character of descriptions need not be very sharp. This is partly due to the fact that on the one hand many proper names have arisen out of descriptions and become proper names as a matter of convention, and on the other hand certain proper names have gradually acquired the function of descriptions and in fact changed into expressions having the character of descriptions. As far as formalized languages are concerned, we are familiar with a number of conceptions based on the total elimination of names. (This procedure incidentally corresponds to the linguistic intuition in natural language by which proper names are replaced by descriptions, identical denotation being preserved.) [9]

(c) *Singular and General Statements*

In the sort of statements which the various branches of science use to express their findings etc., we encounter in addition to words having the character of descriptions, words having the character of names. Let us compare a few statements of which the first group are from physics and the second from history:

(a) 'If the temperature of a given mass of gas remains constant, pressure is inversely proportional to volume.'

'Metal bodies increase in volume under an increase in temperature.'

'Michelson's experiment has shown that either the ether is completely held down by the earth's atmosphere, which, however, is in conflict with Fizeau's experiment, or that our ideas of the ether as a medium which may serve as a reference system are incorrect.'

(b) 'The charter of Charles' University was granted on April 7th, 1348.'

'John Hus was burnt at the stake as a heretic on July 6th, 1415.'

'The independence of the Czechoslovak Republic was proclaimed on October 28th, 1918.'

The superficial comparison of these two groups of statements might lead us to the conclusion that the first operates mainly with descriptions ('metal bodies', 'increase in temperature', 'volume', 'gas', 'ether', 'the earth's atmosphere', etc.), and the second with names ('John Hus', 'the Czechoslovak Republic') or with expressions having the nature of individual descriptions with quite evident singular reference ('the charter of Charles' University'). However, closer comparison reveals that the first group does also operate with individual descriptions or expressions having an analogous position ('Michelson's experiment', 'Fizeau's experiment').

A more detailed analysis of the statements which the various branches of science use to express the results of their activity makes it quite clear that the empirical and experimental sciences cannot be classified purely and simply in accordance as their statements operate with names or descriptions. Nor is it acceptable to classify the branches of science according to whether they use statements operating with names or individual descriptions or statements operating with descriptions alone. One result of this sort of division is the neo-Kantian conception of two groups of sciences, which distinguishes those called generalizing or nomothetic sciences and those called ideographic sciences. The former group includes the majority of the natural sciences, and the latter group includes first and foremost history, a science which describes unique and unrepeatable facts (or, as Ranke put it, it describes "*wie es eigentlich gewesen ist*"). It can, however, be easily demonstrated that the natural sciences must also sometimes describe unique and unrepeatable facts such as, for example, the Tungus meteorite. And on the other hand history – usually on the basis of simple inductive procedures – also makes generalizing conclusions such as conclusions concerning a class of phenomena in a certain period of history, which are statements that by their very nature are not unlike statistical laws.

Thus the occurrence of descriptions or names in the statements of science cannot be taken as a reliable criterion for classifying the latter. In the majority of attempts so far made at such a classification attention has been paid to how the statements can be verified and to whether they are assignable to a single fact (in which case we speak of *singular statements*) or to a certain class of facts (when we speak of *general statements*).

Leaving aside all statements the truth value of which can be decided on the basis of semantic rules – for both the improper and proper signs which occur in these statements (such statements are usually characterized as analytical statements[10]) – as also all contradictory statements, we are left with statements whose truth value must be verified *a posteriori*. The precise delimination of these statements is still a subject of argument and debate[11] and it is not the aim of this work to settle the question. It should be added that the problem of demarcation is more frequently discussed in connection with questions of the fundamentals of logic and mathematics and it is indeed a matter of greater import to these branches of science than to the empirical sciences. From the point of view of the occurrence of names and descriptions the following distinctions should be made:

(a) singular statements with names (e.g. "Karel Čapek was the author of 'RUR'", 'Charles IV founded a university in Prague in 1348');

(b) singular statements with individual descriptions (e.g. 'The present French president is not a soldier', "The author of 'Waverley' became a famous writer of repute");

(c) singular statements with descriptions and words having an ostensive function (e.g. 'The results of the experiment on this apparatus were negative', 'The analysis submitted has established the presence of chlorine in this substance');

(d) singular statements with descriptions containing the exact indication of time and space parameters (e.g. 'At time t and in place p the following data on pressure, temperature, humidity, etc., were found');

(e) general statements, i.e. all statements referring to a defined class of facts (e.g. 'Metals expand with increasing temperature'[12]).

The various writings on the subject contain a number of attempts to eliminate names from statements and replace them by descriptions.[13] Russell, in his criticism of such attempts,[14] has drawn attention to the momentous difficulties of such attempts, concluding that we cannot in fact completely eliminate proper names from the language of science and replace them with descriptions containing explicitly determined time-space co-ordinates. It seems evident that we cannot make an indiscriminate assessment of the possibilities of eliminating names and replacing them by descriptions, treating all the branches of science alike. In a number of branches, in particular the theoretical natural sciences, this tendency has been very positive, enabling the transition from the purely descriptive stage of science to that of theoretical analysis. It is also a quite evident tendency in the various branches of physics. Analogous tendencies can also be found in the biological sciences, for example in the efforts to find a general conception of physiology, the morphology of whole groups of species, etc.

On the other hand, however, it should not be overlooked that the elimination of names and their replacement by descriptions would in many fields lead to a purposeless multiplication of expressions. The reasons in favour of the preservation of names in the language of science are mostly of a pragmatic nature. Russell early pointed out the difficulties which would face the historian if he were to have to replace the name "Napoleon" in all statements of history by corresponding descriptions containing time-space determinations: leaving aside the considerable growth in the bulk of texts and the increase in complexity, which conflicts with linguistic intuition, such a change might lead to statements which would lose their original synthetic character and become analytical. Then of course the objection arises as to the justifiability of any such translation. But Russell himself supports the conception that both names and descriptions have sense and denotation. However, as we have already demonstrated, this viewpoint has limitations, which are connected with the difficulties encountered in assigning sense to expressions having the character of proper names. If an individual description used to replace a proper name is to be an expression which preserves not only the denotation but also the sense of the proper name, then we must assign a sense to the proper name too. This, however, may also lead to difficulties, for instance to the necessity of postulating a special entity distinct from the object named (e.g. what has been called an individual concept), thus violating Occam's razor.

2. Predicates

(a) *The Linguistic Form of Predicates*

We have already pointed out that predicates are assignable to properties. As a property always appertains to a certain object or objects, so too a predicate is not an independent expression – it forms a statement only in combination with suitable arguments. This is why it is irrelevant what kind of words of natural language we use.

In the systems of botanical and zoological expressions we often encounter the form where the decisive element is a noun:

'...is a beast of prey',
'...belongs to such and such a species', etc.

Similar cases are those involving the classificatory approach, relations between element and class, subclass and class, etc.

A form which differs from the foregoing as to linguistic forms used is that which we intuitively understand as the assignment of a property. If we say that

'...has a temperature of 4°C' ('...is so many degrees hot'),
'...is soluble in water', etc.,

we usually operate with either nouns or adjectives. We can equally well use verbs, e.g.

'...boils',
'...dissolves', etc.

We likewise encounter the same parts of speech in many-placed predicates, e.g.

'...has a higher specific gravity than ---',
'...is harder than ---',
'...dissolves in ---',
'...under --- conditions changes into -.-.'.

(b) *Interpretations of Predicates*

The selection of nouns, adjectives or verbs[15] for the expression of predicates in natural language is immaterial from the logical or semantic point of view. Indicative of this is the fact that predicates expressed by means of nouns or adjectives can also be easily expressed by verbs, and vice versa, with little or no change to the sense.

The conception of the predicate as a non-independent expression capable of forming a statement only in combination with suitable arguments comes close to Russell's viewpoint. In Russell's conception the predicate is an incomplete symbol, and, according to this conception, incomplete symbols have no meaning in isolation but only in a certain context. This viewpoint, however, requires that something be added, a closer specification: If we draw a distinction between the sense and denotation of expressions then it is evident that the situation is in either case different. If we say that predicates do not have meaning in isolation but only in relation to objects for which we may substitute a variable from the appropriate domain of variability to which the predicates refer (this domain may also be an empty class), then this limitation concerns not sense but denotation: Predicates in themselves do have sense, we are capable of understanding them and replacing them by other, synonymous expressions, etc. So if we consider predicates as incomplete symbols, we must recognize that they have sense, though we cannot talk of the denotation of predicates in as far as the elements of the universe are only things and not properties. Then predicates cannot of themselves denote things. According to this conception predicates denote something (this 'something' being either all things, some things, a single thing or no thing in the given universe) only when combined with arguments in statements having no free variables.

If, of course, we accept a different semantic starting-point, one corresponding to analysis in the shape 'thing–property–relation', it being taken as fact that a thing has certain properties and is in certain relations, then properties or relations can indeed be assigned to predicates.

Another conception of predicates is represented by Carnap's viewpoint, which follows up Frege's dualistic semantics and reconstructs the assignment of meaning along the lines of the name-relation. As a name denotes an object, so too a predicate denotes a class of objects (this class may of course be empty). Carnap's method of extension and intension operates with two kinds of assignment in the case of all expressions where semantization is possible (such expressions are called 'designators' by Carnap): on the one hand the assignment of an abstract object of a conceptual nature (in the cases of the individual of an individual concept, the predicate of a property,[16] and the sentence of a proposition), on the other hand the assignment of no object, one object or a class of objects in the respective universe.[17] Carnap himself does of course stress that, strictly speaking, sentences alone have an independent meaning, other expressions having only a relatively independent meaning. Nevertheless he does assign parallel sense and denotation to these expressions as well.

There are several ways of proceeding in the semantization of expressions which can be reconstructed, in the language of science, as predicates:

(1) The expressions may be defined, or explicated by the use of other, usually more precise, expressions, or they may be translated into other languages in such a way that we assign to them certain equivalents from the vocabulary of those languages, etc. In this case we are treating predicates as signs in the abstract sense (i.e. Carnap's 'signs-designs' or Reichenbach's 'types') and are specifying the sense of the expressions.[18]

(2) If we take predicates, understood as signs in the abstract sense, as components of language with a precisely fixed universe, it naturally becomes a matter of importance what entities we consider as elements of that universe. Carnap's conception of the denotation (extension) of predicates follows from the assumption that a universe is made up of individual objects. Classes of objects are also an equally legitimate element of Carnap's universe. If we are justified in taking predicates as relatively independent components of the language of science, then (still acknowledging classes as an equally legitimate element of a universe as individual objects) we can assign classes to predicates as their denotation (extension). Thus Carnap's method of extension and intension is justified only on these assumptions, though with the further addition that predicates are not the names of the corresponding classes when the predicates come to be actually used in statements, i.e. that they are not the names of the respective objects in the same sense that a proper name, used in a statement having the character of an actual utterance, is the name of a certain object, person or thing.

(3) If we take predicates as signs in the concrete sense (i.e. as Carnap's 'signs-events' or Reichenbach's 'tokens'), we cannot overlook the context given by actual utterances. This is the case when, for example, we use predicates to express the results of certain scientific procedures – measuring, experiment, confirmation of hypotheses and so forth. In this case predicates are 'incomplete symbols' in Russell's sense, and their semantization is inconceivable without the appropriate context. If we wish to speak here of the sense and denotation of predicates (the one or the other coming more to the fore depending on the context), then this can only mean the sense and denotation of a predicate with respect to the sentence context used and to the precisely fixed universe to which the respective statements, having the character of utterances, refer.

Having reached this situation we should now note a further substantial difference between procedure (1) and procedures (2) and (3): It is a familiar fact that Carnap characterized sense, or the intension of a predicate, as a property. This term corresponds, however, to constructive objects such as Carnap's 'individual concept' or 'proposition'. But if we say that a predicate, used in a certain sentence context having the character of an actual utterance, is assigned to the properties of objects, to their relations, etc., then we are not thinking of a property as a constructive object but properties or relations which we are capable of establishing and distinguishing objectively in the given universe and with respect to the given objects.

Thus when interpreting predicates in the language of science it must be quite clear whether we take predicates as signs in the abstract sense or as signs in the concrete sense (in the latter case the respective context must not be overlooked) and it is also important to know how the given universe has been defined.

(c) *A Criticism of the Nominalistic Interpretation of Predicates*

Our taking predicates as signs in the abstract sense or as signs in a concrete sentence context is also an important factor when we come to assess whether the familiar nominalistic criticism of the hypostatization of the objects which correspond to predicates is justified. Let us assume, for the purposes of illustration, a universe comprising a finite class of physical objects capable of being distinguished by size (i.e. by the two-placed relation $\mathrm{R}x, y$ which we interpret as 'x is greater than y'), by colour P_1 (i.e. $\mathrm{P}_1 x$ is interpreted as 'x is black') and by a further property P_2 (where $\mathrm{P}_2 x$ may be interpreted as 'x is a metal'). In such a situation the adherents of extreme nominalism usually point to the fact that in the given universe there is nothing which would correspond to the two-placed predicate 'to be greater'. There merely exist objects of some of which we can say that they are greater than another object. If this object is b, then all we can say is

$$(\exists x)\ \mathrm{R}x, b \quad (x \text{ is identical with } a).$$

Analogous arguments might be advanced in the case of one-placed predicates, namely that there is nothing in the given universe to correspond to the predicate 'to be black' or to 'to be a metal', but there may exist such objects that

> $(\exists x)\ P_1x$ (x is identical with a) or $(\exists x)\ P_2x$ (x is identical with a) or possibly $(\exists x)\ P_1x\ .\ P_2x$ (x is identical with a).

This is nominalism very strictly understood and it might be formulated thus: Only those entities which can be included among the values of variables such that statements concerning these entities are true, may be acknowledged as existing. Since predicates do not belong among the values of variables, they cannot be considered as expressions to which existing entities can be assigned but as expressions having the character of improper signs or syncategorematic words. It is obvious that criteria for acknowledging the existence of entities conceived on this basis (these criteria are generally described as *ontic commitments*) transfer the responsibility for acknowledging the entities onto the linguistic means selected. If we select, for instance, a language of a higher order which introduces variables for predicates,[18] we must acknowledge the existence of entities which, in the given universe, would correspond to the predicates.

The foregoing formulation for the acknowledgment of the existence of entities is of course too general and indefinite. It was for this reason that Quine, in his formulation of ontic commitments,[19] transferred responsibility for acknowledging the existence of entities onto some theory or other: indeed placing the question of ontic commitments outside the framework of a theory is unjustified. This relativization of ontic commitments can be made more explicit: it cannot be agreed that responsibility for acknowledgment of the existence of entities be transferred onto just any theory[20] or just any conception of the universe. In so far as this responsibility is transferred onto a certain theory, it must be required that this theory be a qualified theory corresponding to the respective demands of the highest level of science attained, which means that this criterion is not relativized but also historicized.

A further step in making more precise the conception of ontic commitments might be based on analyzing the whole problem from the point of view of the communication model of science as described earlier and involving blocks A, B and C.

It is obvious that the nominalistic conception of the criteria for the acknowledgments of the existence of entities transfers this responsibility entirely onto block C, i.e. the language of science, or a certain theory formulated in that language. However, it is scarcely possible to accept such a procedure as satisfactory. Moreover it does not correspond to the way in which entities are reconstructed in science, for example in the domain of the physics of elementary particles and in other natural sciences. This responsibility simply cannot be narrowed down to block C and its components since the selection of linguistic means and the construction of a certain theory must always correspond to blocks A and B and their interrelations. If we take block A, i.e. a certain class of objectively existing entities, as our starting-point, we must respect the fact that these entities have certain objective properties, form certain structures, etc. Which of these properties and relations are distinguishable depends not only on the nature of the properties and relations themselves but also on the properties of block B, for instance on the possibilities of distinguishing and registering these properties, etc.

Thus from the point of view of the communication model of science, responsibility for ontic commitments cannot be placed exclusively on block C (language, theory): it is necessarily bound to the mutual links between all three blocks. This is of course not to say that the responsibility of a qualified theory is suspended, least of all when adequately confirmed by the present stage in the development of observation, measuring, experiments and practical applications. The selection of a more or less suitable block B and C with respect to block A may however always bring attention to bear on certain elements of block A, their properties and relations. In this respect the selection may affect what amounts to a kind of relative apriorism, the formation of relatively *a priori* selection criteria in respect to block A.

The acknowledgment of certain entities as elements of block A leads to the necessity of speaking of one, none, some or all of them and so on. This requires as a rule that variables be introduced for the expressions to which these entities are assignable. In this way the requirement is maintained that we acknowledge as existing such entities the verbal signs of which can be included among the values of variables so that statements about these entities are true. It should finally be noted that this criterion for the acknowledgment of entities is not a starting-point for a conception of the ontic viewpoint, as is the case with the majority of nominalists, but, on the contrary, is a consequence of the acceptance of a certain conception of a given universe.

(d) *'Primary' and 'Secondary' Qualities*

The problem of criteria for acknowledging the existence of entities and the whole problem of ontic commitments may be taken as the starting-point for a new interpretation of Locke's traditional distinction of primary and secondary qualities. The well-known section of the fourth book of Locke's *Essay Concerning the Human Understanding* distinguishes two kinds of qualities (which in the light of the foregoing we consider as predicates): primary and secondary. The former are said to be objective, 'existing in nature'. The latter are subjective or, which amounts to the same thing, exist only in the mind. Locke also defines secondary qualities as powers which produce certain kinds of percepts. From the point of view of our communication model, Locke's differentiation might be interpreted as follows: primary qualities belong to block A, secondary qualities to block B. Analyzing in greater detail what Locke gives as examples of primary and secondary qualities, we find that responsibility for the recognition of objectivity is transferred onto a certain theory, here Newtonian physics, primary qualities being given as those used to describe material phenomena in accordance with the contemporary state of knowledge of Newtonian physics. Thus, for example, velocity, temporal and spatial definitions, etc. are considered as primary. They are after all properties which can be transferred to the properties of the particles of which mass is composed. The mass of a body is given by the number of particles occupying a certain volume. This number is independent of the observer and the presence or absence of an observer has no effect on it whatsoever. A substantially different situation is that of the secondary qualities, whose dependence on the properties and abilities of the observer has regularly been demonstrated on the example of colours.

Locke's epistemic dualism may be criticised from various angles. The most frequent critical procedure has been to accept his starting-point and original characterizations of the qualities and then to prove either that all qualities are primary or that they are all secondary. With the discovery of the physical bases of optical events and the perception of colours it has become relatively easy to prove the primariness of colour qualities. On the other hand some of the subjective-idealist movements have endeavoured to reduce all qualities to something which would correspond to Locke's secondary qualities. A typical example of this is homocentrically interpreted operationalism which interprets even such properties as length – which for Locke were beyond all shadow of doubt primary – as products of the measuring operations of the inquiring subject.

Another method used to criticise Locke's epistemic dualism [21] works with the differentiation of measurable and (hitherto) unmeasurable properties: it can be shown that Locke considered as primary qualities those which can be expressed quantitatively by measuring. In contrast, secondary qualities were those which in his day could not be expressed in measurable quantities. He thus assumed a constant system of invariable standards for measuring, and in accordance with this system, as available in his day, he distinguished between primary (measurable) and secondary (unmeasurable) qualities. This conception of the criterion for distinguishing between primary and secondary qualities lends itself fairly readily to criticism: it can be shown that one and the same quality (e.g. optical and acoustic properties), at one time unmeasurable, is at another time quite easily measurable and expressible quantitatively.

Other critical comments on Locke's epistemic dualism might be raised from the point of view of the formulation of ontic commitments which transfer responsibility for acknowledging the existence of entities onto a certain theory. In the present case there is little difficulty in discovering that the theory which holds responsibility for recognizing the primariness of qualities is Newtonian physics, in which case it must be acknowledged that as a criterion it was limited historically and has long since been overtaken.

Critical analysis of the traditional distinction of primary and secondary qualities has shown that this distinction, which was conditioned by the contemporary state of knowledge, in particular by the level attained in physics at that date, cannot be upheld from the point of view of the present-day level of science. Moreover, the criteria which led to the distinction cannot be considered as absolute, being conditioned historically for example, by the possibilities then available for determining the individual qualities quantitatively. For this reason the methodology of the experimental and empirical sciences no longer makes use of this distinction in our day.

3. The classification of predicates in the language of science: qualitative, comparative and quantitative predicates

In handbooks dealing with the methodology of science we come across chapters dealing with the classification of various types of concepts. As a rule, however, this does not mean the classification of all the components in the vocabulary of the language of science, but only of those expressions which have the character of predicates. From this it follows that the whole domain of improper signs is left to one side, and that, practically speaking, of the proper signs only predicates are taken into account. These restrictions incidentally also apply in the case of certain of the earlier differentiations of concepts, for instance Cassirer's distinction between what he calls substantial and relational concepts.[22] Similarly a number of works dealing with the formation of concepts[23] and their classification in the language of the empirical sciences are essentially confined to the classification of predicates. So whenever in what follows we speak of classificatory or qualitative concepts, comparative concepts, quantitative concepts, empirical, dispositional and theoretical concepts, what we have in mind are those expressions which have the character of predicates.

What then presents itself as the simplest classification of predicates, and in fact it is a formal classification, is their division according to the number of arguments into one-, two-, three-placed, etc. At the same time it is not possible in theory to limit the number of arguments, but in practice we generally confine ourselves in the language of science to predicates with one or only very few arguments. The reasons here are essentially of a practical nature, but they are also supported by the actual nature of those expressions which have the character of predicates in ordinary language, namely nouns, adjectives and verbs. Although it is possible, especially in the case of verbs, to reconstruct these expressions as predicates with more than three arguments, these are in fact relatively restricted cases, the majority of predicates being indeed one-, two-, and, in a small number of cases, three-placed.

The differentiation of predicates according to the number of their arguments contrasts with another differentiation in which the number of arguments still has a part to play but which does not completely coincide with the differentiation merely by the number of arguments alone. From the methodological point of view, it is of greater importance to distinguish predicates according to the role they have with respect to the objects of a given universe. From this point of view we may then distinguish

(a) qualitative (known also as classificatory) predicates,
(b) comparative predicates,
(c) quantitative predicates.

The simplest way of characterizing the role of the various kinds of predicates with respect to the objects of a given universe is as follows: Qualitative predicates give a simple definition of the properties of individual objects. For this reason they are capable of realizing only the *decomposition* of the given universe down to the class of objects having the respective property, but not a class complementary to this one. For these reasons qualitative predicates represent the starting-point for the decomposition of the given universe and so for the classification of the objects in that universe. Comparative predicates are capable of realizing the *organization* of the objects of the universe. This assumes that these predicates express relations, but only such relations as can ensure organization. This is why comparative predicates are not identical with the class of many-placed predicates, being merely a sub-class of the latter. In the case of quantitative predicates what is required is quantitative determination, i.e. numerical characterizations of the properties of objects expressed by a single number, a pair of numbers, three numbers, etc.

(a) *Qualitative Predicates*

When chemistry divides substances into metals and non-metals, when physics distinguishes between the gaseous, liquid and solid states of substances, and when in psychology we divide personalities by temperament into choleric, sanguine, melancholic and phlegmatic, we are in fact giving certain qualitative characterizations of the objects of the given universe and so too a classification of these objects. For these reasons qualitative predicates are usually characterized as classificatory concepts. The bond between qualitative concepts and classification is not always clear: in the case of predicates such as 'red', 'translucent', 'flexible', etc. there seems to be no more than merely the qualitative determination; in fact even these predicates may be used in the decomposition of a universe and so too in a classification, at least by distinguishing classes of objects having the relevant property, and the complement of each class.

We most frequently assume qualitative predicates to be one-placed, but this is far from being the rule. Many-placed qualitative predicates are also perfectly common, for example '*x* is a friend of *y*', '*x* is near *y*', '*x* is parallel to *y*', etc. Also many-placed predicates may be the starting-point for the decomposition of the objects in a given universe, though of course the decomposition is ambiguous. If, for example, we have the two-placed predicate '*x* is a friend of *y*', or '*x* is parallel to *y*', and if the universe is comprised of the class of people in a certain town or that of straight lines in a given plane respectively, we may proceed in just the same way as with one-placed predicates, providing we replace one of the variables by a constant. This practically converts the two-placed predicate into a one-placed and we proceed with the decomposition according to this one-placed predicate. The other way of decomposing the universe is to divide all pairs of persons and all pairs of straight lines into those pairs standing in the mutual relation of friendship and those pairs of straight lines which are parallel on the one hand, and the respective complementary classes on the other hand.

Qualitative predicates as a rule represent the most elementary characterization of objects, which then allows of being made more explicit and even of being expressed quantitatively. The same also goes for many classifications based on qualitative predicates: the Beaufort scale of wind velocity was originally based on the purely qualitative determination of the various kinds of wind. This scale can, however, be expressed in some cases quantitatively *ex post*, by, for example, measuring wind velocity in metres per second. In the same way the periodic table of the elements, the anthropological scale of human skull types, and others, can all be expressed quantitatively. In other cases no such quantitative expression of qualitative differences has been arrived at, or else the quantitative determinations so far available to us are not capable of expressing all qualitative differences. This is the case with, for instance, a number of taxonomic classifications in zoology, botany and elsewhere. In spite of the fact that science will always, quite naturally, endeavour to draw distinctions on the basis of quantitative determination, qualitative predicates are still of importance to science in our day. In a whole range of branches of science qualitative determinations are quite indispensable, not merely because the quantification of these determinations would be impossible, but chiefly because the ways of quantification which mankind has at its disposal to date are incapable of expressing without remainder all that is expressed in the qualitative determinations.

In connection with the role of qualitative predicates as starting-points of classifications, a special group is sometimes made of predicates capable of ensuring 'natural classification' as opposed to those which merely make 'artificial classification' possible. However, the difference between natural and artificial classification is merely relativized and, moreover, highly vague. If we were to classify all animals into various classes by weight or size, we should also arrive at a classification of a sort. Such a classification, however, realizes the decomposition of the respective universe according to what – in the given situation – are irrelevant properties. In contrast to this, familiar taxonomic classifications are said to be natural. Natural classifications are then said to follow relevant characteristics of the objects of a given universe. The notion of the 'relevant characteristics' of objects is vague, to say the least, and its definition depends first and foremost on circumstances of a pragmatic nature: that which is for certain purposes declared to be a relevant characteristic need not be such for other purposes. Some sort of relative justification for natural and artificial classification might be discerned in the fact that in the first case the classification respects what might be called a *resemblance-class*. The latter is determined by an entire group of predicates, so that if an element of a resemblance-class differs from other elements of a given universe in that it has the property P_1, then it also has the properties $P_2,\ldots, P_n$, where $\{P_1, P_2,\ldots, P_n\}$ denotes the collection of properties which characterize the resemblance-class. Thus we can say that any object having property P_1 also has the properties denoted $P_2,\ldots, P_n$. We are usually inclined to take the most conspicuous determination of the collection $\{P_1, P_2,\ldots, P_n\}$ as the relevant characteristic, in spite of dealing in fact with a resemblance-class. For this reason all the categories which emerge in taxonomic classification, i.e. order, genus, family, species, etc., may be considered as resemblance-classes which can only be characterized by a certain group of predicates between which there are always certain fixed dependence relations.

Classifications in the empirical and experimental sciences have followed what we have described as resemblance-classes. We usually require of a classification that it decompose a given universe in such a way as to place objects which are similar in some way in the various sub-classes. Intuitively we usually understand a classification as a decomposition of the given universe on the basis of those properties which are denoted by one-placed predicates. This does not of course render impossible the kind of classification which includes ordered pairs, triplets, ... *n*-tuples of objects in individual sub-classes.

For the sake of simplicity in the analysis to follow we shall be confining ourselves to one-placed predicates, which is incidentally quite in accord with the intuitive understanding of classification. We shall further assume that we are able to place any object either in the class $(\hat{x})\, P_i x$ or in the class $(\hat{x}) \sim P_i x$, no object remaining unplaced and no doubts arising as to the correct placing. In other words we assume that the principle of 'tertium non datur' holds and that the predicates which lie at the basis of the decomposition are not vague.[24]

The concept 'resemblance-class' suggests that in classifications in the empirical and experimental sciences we assume that the various predicates are dependent on each other. This means that having decided on the assignment of one property we are then capable of deciding on the assignment of others. Indeed one of the major roles of any classification is to make such decision-making possible. Thus if the interdependences of the various predicates are known, in the form, for example, of meaning postulates, then simply on the basis of the approach to the classification certain other properties can be foreseen. For example, if we characterize a resemblance-class by the group of predicates $\{P_1, P_2, P_3\}$ and if we know that $(\forall x)\,(P_1 x \rightarrow P_2 x)$ and $(\forall x)\,(P_1 x \rightarrow P_3 x)$, then the discovery that a given object belongs to a given class because it has the property denoted P_1 warrants the conclusion that it will also have the properties denoted P_2 and P_3. If, moreover, we are capable of reconstructing a resemblance-class of this kind as a possible sub-class in a given universe, we can conclude in advance what will be the properties of the objects which belong to this sub-class irrespective of whether any such objects have been established empirically or not. A classical example of the kind of classification that enables us to foresee the properties of objects if we know to which class they belong is Mendeleyev's periodic table of the elements.

A classification of the objects of a universe which enables us to decide on the properties of those objects (and also foresee these properties) in accordance with our forehand knowledge of resemblance-classes and of the interdependences of the properties of these classes, is in fact a higher type of classification. However, it is not difficult to imagine lower types of classification where we do not know these interdependences, or where at least we do not take them into consideration. In this case the decomposition of a given universe may be carried out in accordance with the method of state description and structure description. If $\mathscr{X}$ is a universe where ($x \in \mathscr{X}$) and we have at our disposal π one-placed predicates (i.e. $P_1, P_2, \ldots, P_\pi$) for its decomposition, we may create a total of 2^π subclasses. This decomposition may also be conceived in such a way that on the basis of given atomic predicates, i.e. on the basis of $\{P_1, P_2, \ldots, P_\pi\}$, we constitute molecular predicates $\{Q_1, Q_2, \ldots, Q_\alpha\}$, where $\alpha = 2^\pi$ and Q_i is defined as a conjunction of atomic predicates or their negation such that each P_i occurs in the definens only once and the possibility of the occurrence $P_i \,.\, \sim P_i$ is excluded. Taking for example the triplet of atomic predicates $\{P_1, P_2, P_3\}$ we may form 2^3, i.e. 8, Q-predicates and thus 8 sub-classes according to the plan (each Q_i being capable of forming a class $(\hat{x})\, Q_i x$):

$$
\begin{aligned}
Q_1 &= df \quad P_1 \,.\, \quad P_2 \,.\, \quad P_3\,,\\
Q_2 &= df \quad P_1 \,.\, \quad P_2 \,.\, \sim P_3\,,\\
Q_3 &= df \quad P_1 \,.\, \sim P_2 \,.\, \quad P_3\,,\\
Q_4 &= df \quad P_1 \,.\, \sim P_2 \,.\, \sim P_3\,,\\
Q_5 &= df \sim P_1 \,.\, \quad P_2 \,.\, \quad P_3\,,\\
Q_6 &= df \sim P_1 \,.\, \quad P_2 \,.\, \sim P_3\,,\\
Q_7 &= df \sim P_1 \,.\, \sim P_2 \,.\, \quad P_3\,,\\
Q_8 &= df \sim P_1 \,.\, \sim P_2 \,.\, \sim P_3\,.
\end{aligned}
$$

The use of molecular predicates affords certain advantages: It is possible, for example, to select a disposition of the classes $(\hat{x})\, Q_1 x$, $(\hat{x})\, Q_2 x, \ldots$, etc., so that the classification suggested guarantees an approximation to the normal disposition. Moreover what we actually require of the classification, i.e. that the various classes are disjunct, is guaranteed.

It is natural that the method of classification suggested is suitable where we take no account of any interdependences between the atomic predicates or wherever any such interdependences are still unknown.

Nor is it impossible that when certain interdependences have been established, certain of the possible Q-classes may turn out to be empty.

Classification based on the conjunctive connection of all atomic predicates or their negations has certain advantages over traditional modes of classification, such as dichotomous classification. The application of the individual predicates in a dichotomous classification can be illustrated in the shape of the following diagram, where a universe $\mathscr{X}$ ($x \in \mathscr{X}$) is decomposed into sub-classes by the use of the individual predicates (see Figure 3).

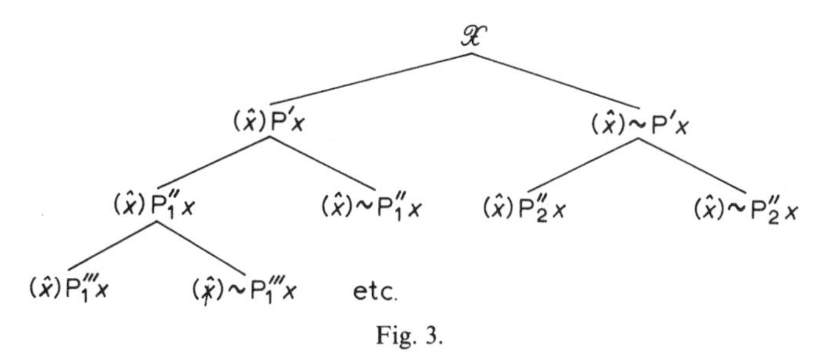

Fig. 3.

A simple comparison of this and the previous method of classifying a given universe by the use of a (finite) class of classificatory predicates makes clear the advantage of that mode of classification which relies on the conjunctive connection of all atomic predicates or their negations. In a decomposition which makes use of the construction of Q-predicates on the basis of π atomic predicates we obtain a total of 2^{π} Q-predicates and so too 2^{π} sub-classes. In contrast the second way gives only 2π sub-dependences between the individual predicates, such as that of incompatibility between P''_1 and $\sim P'$, i.e.

$$(\forall x)\,(P''_1 x \mid \sim P'_{x}),$$

incompatibility between P''_1 and P''_2, i.e.

$$(\forall x)\,(P''_1 x \mid P''_2 x) \text{ etc.}$$

It is evident that in this way certain interdependences between predicates are introduced implicitly, though it is always necessary to give them precise formulation and to see that this formulation is as exhaustive as possible, which of course makes it all the more difficult to carry out the whole classification.

Even greater difficulties arise with classifications where each class, or sub-class, decomposes not into two, but into three, four or more disjunctive sub-classes. In this case the network of interdependences becomes even more complicated and even less readily surveyable.

(b) *Comparative Predicates*

Comparative predicates, known also as order or topological concepts, always represent relations and therefore have the form of at least two-placed predicates. We should also mention here the relation between qualitative and comparative concepts: with qualitative concepts the assumption behind the assignment of properties is that of decision-making of the type *either – or*. From the extensional point of view this means that an object a ($a \in \mathscr{X}$) is either an element of the class $(\hat{x})\, P_i x$ or an element of the class $(\hat{x}) \sim P_i x$.

With comparative predicates we assume that a certain property does belong to an object but that the assignment of the property can be realized in different *degrees*. Here then decision-making is of the type *more or less*. Comparative predicates are connected as a rule with certain qualitative predicates, the qualitative predicates 'heavy', 'resilient', 'warm', for example, corresponding to the comparative predicates 'heavier', 'more resilient', 'warmer', etc. It is not absolutely essential here to assume that qualitative predicates will have the attributive form since comparative predicates can also be constructed to correspond to qualitative predicates which, from the linguistic point of view, do not have the attributive form, e.g.

> 'to have a volume of' – 'to have a greater volume than',
> 'to be a conductor of electricity' – 'to be a better conductor of electricity than'

etc.

The necessity of having comparative predicates and so the possibility of assigning a given property in varying degrees usually results from the fact that a class $(\hat{x})\, P_i x$ and its complement $(\hat{x}) \sim P_i x$ (assuming P_i to be a qualitative predicate) are not homogeneous classes. One aspect of this is that it is difficult to decide on the precise boundary between $(\hat{x})\, P_i x$ and $(\hat{x}) \sim P_i x$, so that the degree of certainty as to the inclusion of various elements in the two classes may vary. In other words if two or more objects are elements of one and the same class, and if they share a certain property, there may be differences in the degree to which they have this property. Comparative concepts are introduced so that the elements to which we assign a certain property can be *ordered* according to the degree in which they have the property. For example, if the objects $a_1, a_2, \ldots a_n$ are conductors of electricity, then, provided we are able in the case of each pair of objects to decide that they have the property 'to be a conductor of electricity' to a greater, lesser or equal degree, we can order the class $(\hat{x})\ P_1 x$ (i.e. the class of all conductors of electricity).

The possibility of ordering, which is the basic function of comparative concepts, results then from the possibility of comparison. This also means that a comparative predicate must have more arguments than the corresponding qualitative predicate. If, for example, P_1 is a one-placed predicate 'to be hard', then in the case of the pair a_1 and a_2 we can state whether they are more or less hard than each other or equally hard. Thus the counterpart of the one-placed qualitative predicate 'x is hard' is the two-placed comparative predicate 'x is harder than y'. If a qualitative concept is a two-placed predicate (e.g. 'x is far from y'), then the corresponding comparative predicate is four-placed ('x is further from y than u is from v'). Many-placed comparative predicates are usually reduced for expediency to predicates with less arguments, which can be done in various ways. In the case of a four-argument comparative predicate there are two ways of converting it into a two-argument predicate. Either we compare directly two pairs of objects so that the predicate by which we make the comparison is in fact two-placed. For example, having introduced the concept of distance we may say that the distance between one pair is greater than that between the other pair. Or we may replace two of the arguments by constants so that the resulting predicate will be only two-placed. (In the above example of a four-placed predicate we might replace the variables u and v by certain objects, e.g. the objects 'London' and 'Paris', and obtain the predicate 'x is further from y than London is from Paris'.)[25] In a way analogous to this, two-placed comparative predicates may also be reduced to one-placed by replacing one of the variables by a constant. Taking as an example the predicate 'x is harder than y', we may replace the variable by the constant 'quartz' to give the qualitative predicate 'x is harder than quartz'. In a universe of ten objects (those which serve as degrees on Mohs' scale of hardness), namely talc, gypsum, calcite, fluorite, apatite, orthoclase, quartz, topaz, corundum and diamond, we discover that the first six items are elements of the class of objects which are not harder than quartz, while the last three are elements of the complementary class.

In the remaining part of the analysis of the basic properties of comparative predicates we shall be confining ourselves merely to those comparative predicates which have the character of two-placed predicates. It must be stressed that not all two-placed predicates have the character of comparative predicates: for instance the predicates 'to be similar to', 'to be parallel to', 'to be perpendicular to', 'to be identical to', etc. are also two-placed predicates though they are incapable of ordering the objects of a given universe.

We shall consider as comparative predicates – the class of which we denote $\mathscr{A}$ with elements $A_1, A_2, \ldots, A_n$ – only such two-placed predicates (the class of which we denote $\mathscr{P}^{(2)}$ with elements $P_1^{(2)}, P_2^{(2)}, \ldots, P_n^{(2)}$, it being also true that $\mathscr{A} \subset \mathscr{P}^{(2)}$) as are capable of ensuring *the complete ordering* of all the objects of $\mathscr{X}$. This means that these predicates must be *irreflexive*, *transitive* and therefore *asymmetrical*. It then holds (A_i being any comparative predicate) that

$$(\forall x) \sim A_i x, x,$$
$$(\forall x)(\forall y)(\forall z)(A_i x, y \,.\, A_i y, z \rightarrow A_i x, z),$$
$$(\forall x)(\forall y)(A_i x, y \rightarrow \sim A_i y, x).$$

If A_1 is the comparative predicate 'to be harder than' then we are in a position to decide whether either one of two solid bodies is harder than the other, or, which is a further possibility, whether they are equally hard. Thus it becomes evident that in the case of comparative predicates the disjunction $A_i x, y \vee A_i y, x$ need not exhaust all eventualities. Here we have, and in practice this is of great importance, a third possibility whereby the two objects cannot be differentiated with respect to the given comparative predicate. In this case we say that the pair of objects *coincide* with respect to the given comparative predicate. The concept of coincidence, which must always be relativized with respect to a given comparative predicate, should likewise be treated as a two-placed predicate. If we express coincidence respective to A_i in the form C_{A_i}, then it holds that

$$(\forall x)(\forall y)(A_i x, y \vee A_i y, x \vee C_{A_i} x, y),$$

in which case we say that the comparative predicate is *C-connected.*

If any two objects do not coincide with respect to A_i, we are capable of deciding whether $A_i x, y$ or $A_i y, x$, i.e.

$$(\forall x)(\forall y)[\sim C_{A_i} x, y \rightarrow (A_i x, y \vee A_i y, x)].$$

The concept of coincidence also has important properties: it is transitive, symmetrical and reflexive. Thus it holds that:

$$(\forall x)(\forall y)(\forall z)(C_{A_i} x, y . C_{A_i} y, z \rightarrow C_{A_i} x, z),$$
$$(\forall x)(\forall y)(C_{A_i} x, y \rightarrow C_{A_i} y, x),$$
$$(\forall x)\, C_{A_i} x, x.$$

The pair of concepts A_i and C_{A_i}, i.e. a comparative concept and the corresponding concept of coincidence, determine a quasi-series.[26] Since each comparative predicate has a counterpart in the respective concept of coincidence, it is theoretically possible to set up as many quasi-series in a given universe as there are comparative concepts. In the empirical and experimental sciences it is naturally expedient to limit the number of quasi-series and reduce the number of comparative concepts to as many as will suffice for the various roles associated with comparative predicates. (Here we have in mind, for instance, comparison of the objects of a given universe, the investigation of their similarity, the role of identification, measuring, etc.) Due account should also be taken of the fact that comparative predicates are not independent of each other; it can be shown that there are certain meaning relations between them. One thing that this leads to is that certain quasi-series originating out of different comparative predicates may coincide, different comparative predicates thus leading to the construction of mutually interchangeable or similar quasi-series.

The statement that the pair of concepts A_i and C_{A_i} form a quasi-series in the universe $\mathscr{X}$, must refer to the whole universe: there are some predicates, similar to comparative predicates in that they permit of the assignment of a property in varying degrees, that are capable of ordering only a part of the universe and giving no guarantee of ordering the whole of it. One example is the relation 'to be victorious over', which for the participants in a tournament is not an ordering predicate. It can be shown similarly that some apparently comparative predicates (e.g. 'to prefer', 'to be more useful than', etc.) are not transitive over the whole scope of the universe but only over a part. It follows that these only apparently comparative predicates cannot ensure the formation of quasi-series in the whole scope of a universe. In such situations there are two possibilities open to us: on the one hand it is possible to so modify the universe that the predicate in question does ensure its total ordering, or, on the other hand, the predicate must be so modified that it becomes capable of ensuring that the whole universe, and not merely one part, is ordered.

If $\mathscr{X}$ is a universe and $\mathscr{A}$ a finite class of comparative predicates, a finite number of quasi-series can be constructed in $\mathscr{X}$. If $\mathscr{X}$ has a large or confused mass of elements, some of them may be selected as place-markers in a given quasi-series. This makes possible the formation of the simplest scale capable of indicating the place held by any object in the given quasi-series. This is the case with Mohs' scale of hardness, the Beaufort scale of wind velocity, etc. This means that the whole quasi-series is split up into a relatively small number of segments which can, if required, be presented in the form of a sequence of integers, letters of the alphabet, etc. For expression diagrammatically, the universe $\mathscr{X}$, represented as a band, can be split up into a finite number of segments, each segment either coinciding with, or being greater than the respective marker (see Figure 4).

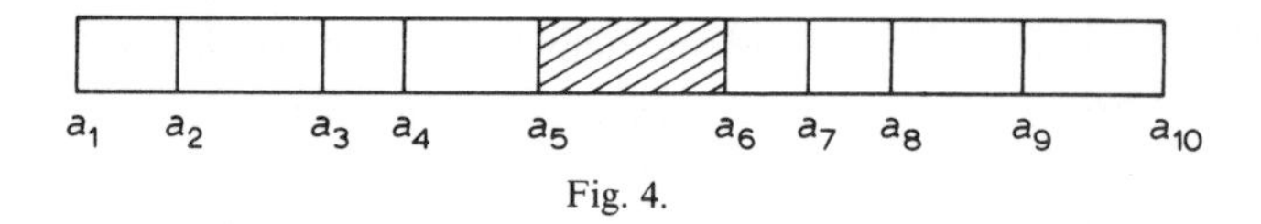

Fig. 4.

What is recorded on the diagram are the positions of the objects which serve as place-markers in the quasi-series.[27] It is then possible in this situation to establish the position of an object in the quasi-series by using qualitative predicates formed from comparative predicates by replacing one of the variables by a constant from the class $\{a_1, a_2, \ldots, a_{10}\}$. If a_1 is talc, a_2 gypsum, a_3 calcite, a_4 fluorite, a_5 apatite, a_6 orthoclase, a_7 quartz, a_8 topaz, a_9 corundum and a_{10} diamond, then objects which are harder than apatite and not harder than orthoclase (such items would belong in the shaded block of the diagram) constitute a class arising from the intersection

$$(\hat{x})\, \mathrm{A}_1 x, a_5 \cap (\hat{x}) \sim \mathrm{A}_1 x, a_6 .$$

It is a familiar fact that 'to be harder than' is determined operatively: any item x is harder than a_i if x scratches a_i but not vice versa. However, of practical importance is the situation where the two items are equally hard. It is therefore expedient to introduce the modified qualitative predicate 'equally hard or harder than a_i' expressed as $A'_1 x, a_i$, where

$$A'_1 x, a_i \equiv A_1 x, a_i \vee C_{A_1} x, a_i .$$

In accordance with this we can mark with greater accuracy a position in the diagram which will include the forward limit (drawn in in full) while not including the far limit (dotted) (see Figure 5).

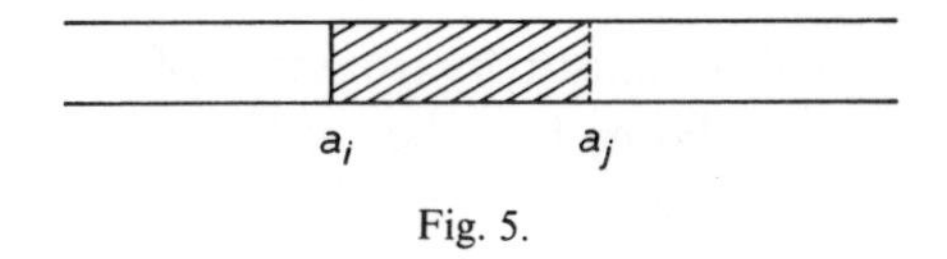

Fig. 5.

Then objects which are as hard as or harder than a_i and simultaneously are not as hard as or harder than a_j constitute the class formed by the intersection

$$(\hat{x})\, A'_1 x, a_i \cap (\hat{x}) \sim A'_1 x, a_j .$$

From these modified qualitative predicates it is possible to make the transition to modified comparative predicates of the type 'in the same measure as or more than', which in the case of our example is 'as hard as or harder than', which we express by the sign $A_i x, y$, where

$$A'_i x, y \equiv A_i x, y \vee C_{A_i} x, y .$$

Modified comparative predicates have certain advantages over normal comparative predicates, though they also have some disadvantages. Modified comparative predicates may be considered as more general in that they include relative coincidence as a borderline case. It is also possible, as follows from the diagram, to determine with greater precision the boundaries of a class of objects having the respective property in an equal or greater measure with respect to a given marker. Another advantage, as Carnap[28] has pointed out, is that on the basis of modified comparative predicates it is possible to define not only comparative predicates of the ordinary kind but also predicates for relative coincidence. This means that

$$A_i x, y = df\ A'_i x, y\,.\sim A'_i y, x\,,$$
$$C_{A_i} x, y = df\ A'_i x, y\,.\,A'_i y, x\,.$$

At the same time, however, the disadvantages of modified comparative predicates should not be overlooked. These predicates represent reflexive relations, so that

$$(\forall x)\ A'x, x\,.$$

From the formal point of view we might presume transitivity here, i.e.

$$(\forall x)\,(\forall y)\,(\forall z)\,(A'x, y\,.\,A'y, z \rightarrow A'x, z)\,,$$

but it is a different kind of transitivity from that of normal comparative concepts, since it can be borne by either of the two distinct aspects of modified comparative predicates, i.e. the aspects of 'in the same measure' and 'in a greater measure'.[29]

On the other hand we cannot prove either that these predicates are symmetrical or that they are asymmetrical, since there is never any guarantee as to which of the two aspects applied in any given situation. This factor considerably complicates work with these predicates in the context of an empirical universe.

(c) *Quantitative Predicates*

From the logical viewpoint, quantitative predicates, also known as metrical or numerical concepts, do not differ from either qualitative or comparative predicates. If we say of a given object that it 'is of the degree of hardness 5' (according to Mohs' scale), 'weighs 5 g', 'is 5 cm long', 'has a velocity of 5 m/s', 'has a consumption of 7 litres of petrol per 100 km', etc., we are in fact operating with one-place predicates. These differ from ordinary qualitative predicates in that the determination of the quality is made using a single real number or a pair, triplet... or n-tuple of real numbers.[30] In this sense a quantitative predicate may be treated as the representation of a qualitative predicate in one or more sets of real numbers. What we have here said on the relation between qualitative and quantitative predicates also applies, with a slight modification, to the relation between comparative and quantitative predicates. Against the comparative predicate 'x is greater than y' we can construct the quantitative predicate 'x is twice as great as y'. From all this it becomes evident that the division into qualitative, comparative and quantitative predicates is not altogether correct as a classification, since not only predicates having originally the character of qualitative predicates but also those having originally that of comparative predicates can be expressed with greater precision quantitatively. It is for this reason that Carnap considers quantitative predicates to be explicata of qualitative and comparative predicates, and also comparative predicates to be explicata of qualitative predicates.[31] It is indeed beyond all question that in the language of science the majority of quantitative predicates can be defined with greater precision and less ambiguity, and that it is possible to eliminate the difficulties with which many qualitative predicates are beset, especially those connected with the possibilities of subjective interpretation, vagueness, ambiguity, etc. In spite of this there can still be found qualitative predicates to which explicata in the shape of quantitative concepts cannot be assigned. These are above all qualitative predicates which express properties not distinguishable as to the degree in which they are present, i.e. those which cannot be ordered on the principle of 'more' or 'less'. Thus the possibility of constructing comparative predicates in respect of given qualitative predicates is an important symptom of possibilities of quantification.

Quantitative predicates denote the quantified properties of objects which are expressed in certain quantities: length in centimetres, metres, kilometres, etc., velocity in units of length per unit of time, resistance in ohms, etc. Thus to the foregoing description of quantitative predicates as a representation of qualitative predicates in one or more sets of real numbers we must now add a certain semantization of these sets: they are the set of metres, grams, ohms, ampères, horsepowers, etc. We need not of course interest ourselves here with the origins of these quantities, the accuracy of their definition and suchlike questions. In literature dealing with the logical and semantic problems of measuring, attention is quite properly drawn to the anthropocentric character of the earliest quantities used – feet, spans, inches, steps and the like. The next stage in making these and similar quantities more precise was the necessary *standardization* of units, which then fixed the convention as to the properties of these quantities. The metric system and other systems of units dependent on it represent the next stage, connected not only with the endeavour to desubjectivize the units but also with that of making them mutually comparable and reducible and of ensuring those properties which we require of measuring, especially those of the metrization of quasi-series and of additivity. These quantities and the systems in which they are involved are always of a conventional nature, so that quantitative predicates represent the kind of determination of objects and their properties which involves, explicitly or implicitly, the acceptance of a certain convention, i.e. the acceptance of a certain system of quantities.

Thus, the quantification of properties which are expressible by quantitative predicates is carried out using elements from a set of real numbers and certain conventional quantities, which are represented by a class of scale factors. Let us denote this class $\mathscr{M}\,[\mathscr{M} = (m_1, m_2 \ldots)]$. Let us now assume that $\mathscr{M}$ is the class of ten substances on Mohs' scale for establishing degrees of hardness, i.e. talc, gypsum ... diamond. If we are then to discover degrees of hardness in a universe $\mathscr{X}$ (where naturally $\mathscr{M} \subset \mathscr{X}$), then a substance x is of hardness $i\,(i = 1, 2, \ldots\ 10)$, if it holds that

$$C_{A_1}x, m_i,$$

where A_1 is the comparative predicate 'to be harder'.[32] In this case the quantification is carried out using elements from the set of integers and the conventional quantities represented by the components of the Mohs scale. One precondition of this procedure, i.e. the assignment of a certain numerical value to a degree of hardness is the *numerization of the quasi-series of class* $\mathscr{M}$, this numerization serving to numerize the elements of $\mathscr{X}$.

This procedure meets only the most elementary demands of the quantification of properties. A higher form of quantifying properties is represented by procedures using the *metrization of the quasi-series of class* $\mathscr{M}$, which assumes that the intervals between each m_i and m_{i+1} can be determined by a general rule.[33] Let the class $\mathscr{M}$ consist of weights of 1 g, 2 g, 3 g, 4 g, ... etc., and let us assume that we are to establish the weight of objects. If A_1 is the comparative predicate 'to be heavier' and C_{A_1} the corresponding predicate of coincidence, then an object under investigation has weight i, if it holds that

$$C_{A_1}x, m_i.$$

This means that if we can assign quantity i to scale-factor m, then we can also assign it to object x.[34] We assume of the majority of quantifiable properties that the quantity i is continuously variable. This means that in a universe $\mathscr{X}$, i may acquire values which are real numbers. On the other hand, a quasi-series constructed in $\mathscr{M}$ may be represented either as a set of integers – as is the case with the Mohs scale but also with weighing using weights of 1 g, 2 g, 3 g, etc. but without the use of smaller weights – or as a set of real numbers. The latter case assumes, however, that the metrization of a quasi-series in the class $\mathscr{M}$ can be represented in a set of real numbers. This representation, however, is never without limitations, which are imposed by the limited degree to which the scale factors can be refined, etc. Since the metrization of a quasi-series in the class $\mathscr{M}$ is always restricted in this way, there are always certain limits to the quantification of predicates or to the accuracy of measurement. In establishing the relation $C_{A_1}x, m_i$ it is moreover necessary to calculate with the fact that throughout the procedure essential to this, the relevant characteristics of both x and m_i remain constant, at least, that is, until the relation is established. Since, however, these essential procedures always take place in time and space, this relative constancy is always limited. Therefore any quantification of the properties of objects is abstracted from the possibility of changes taking place in these properties. It is also abstracted from the possibility that while the relation $C_{A_1}x, m_i$ is being established, the object m_i may act on object x in such a way as to change the properties of x which are to be quantified, or alternatively object x may so affect m_i as to change the properties of m_i which are relevant for the quantification.[35]

The establishment of the relation $C_{A_1}x, m_i$ on the assumption, for example, that A_1 is the comparative predicate 'to be heavier' and that the scale factors are weights graded by one gram, justifies us in stating that an object x has a weight of i grams. Analogous considerations might be applied to the measurement of length in centimetres, capacity in litres, the price of goods in the units of a national currency, etc. Let us then proceed to investigate what it means when an object y is twice as heavy as an object x, or twice as long, or has twice the capacity, price, and so forth. If to an object x we assign a weight represented by the numerical value i (or a length of i cm, a capacity of i litres, etc.), then we may ascribe to the object y a weight (length, capacity, etc., respectively) represented by the numerical value $2i$. Similarly we might investigate the weight of the body which arises out of the physical joining-together of objects x and y, if the weight of x is represented by the numerical value i and that of y by the numerical value j. If it holds that the weight of the object arising out of the combination of objects x and y may be represented by the sum $i+j$, then we say that the quantity which characterizes weight is *additive*. A quantity expressing a property is said to be additive if the physical combination of two or more objects gives rise to an object whose property is represented by a numerical value equal to the sum of the values representing the properties of the original objects.

Weight, capacity, length, price and others are examples of additive quantities.[36] However, other properties, not expressible in terms of additive quantities, can also be quantified. For example, if we take two vessels containing the same liquid but at different temperatures, say i °C and j °C, and then pour the contents of both vessels into a single vessel, the resulting liquid will not have a temperature of $i+j$ °C. Likewise quantities expressing density are also not additive: a substance which arises out of the combination of two other substances of differing density will not be of the density represented by the sum of the numerical values of the original densities.

The distinction between quantities which can be considered as additive and those which cannot is connected with a number of complex problems which have long been the subject of analysis. Campbell[37] was the first to distinguish between extensive properties, or additive quantities, and intensive properties (Campbell originally called them qualities), or non-additive quantities. Extensive properties are those for which there exist empirical operations similar to the arithmetical operation of addition. Intensive properties are those for which no such operation exists.[38] A viewpoint not unlike that of Campbell's differentiation of extensive and intensive properties is that held by Cohen and Nagel[39] who emphasize that fundamental measuring is only possible on the basis of extensive properties.

A different attitude to the distinction between extensive and intensive properties is taken by Hempel.[40] Above all he does not consider this to be an absolute difference, and relativizes it with respect to a certain physical operation of combination: the resistance of conductors in electricity is additive, but only if these are ordered in series. This same property, which in this way appears as extensive, ceases to be additive if the conductors are ordered in parallel. The length of bars is additive if they are placed end to end and if the space between them approaches zero. If, however, we join the bars together with a different material, or in the case of welding metal bars together, we must always allow for a difference between the final value and the simple sum of the two lengths. In this way additivity is relativized to a precisely specified mode of (physical) combination or to a precisely specified operation of combining or accreting the original objects. This also means that with a change in this relativization, some of the quantities which were counted on as additive may become non-additive. This circumstance may also be a source of error in quantification.

Hempel, in his relativization of the differences between extensive and intensive properties, goes even further. Modes of the physical combination of objects, or operations for their combination, are naturally not selected at random. Their selection depends on the overall theoretical conception of the given universe. It is for this reason that Hempel places the requirement that the mode of combination or the operation of combination of objects, together with the corresponding comparative predicates and predicates of relative coincidence, be tied to what he calls a 'simple and fruitful theory'.[41] For example, in measuring hardness in mineralogy, we still do not possess any such theory nor any way of combining minerals which would ensure that the property be extensive.

Any operation of the (physical) joining-together or combination of original objects which would be capable of ensuring that quantities be additive, must satisfy certain conditions analogous to those for counting in number theory. We shall express the properties of this operation by using the predicates P_1, A_1 and C_{A_1}, predicate P_1 being a qualitative predicate which allows of the given property being assigned in varying degrees, A_1 being the corresponding comparative predicate, and C_{A_1} being the respective predicate for coincidence with respect to A_1. (P_1 might be, for example, the property of conductors of offering resistance to an electric current, A_1 is then 'to offer greater resistance than', and C_{A_1} is 'to offer the same resistance as'.) Let us denote the operation of (physical) combination by the sign $\circ$. We must assume that the application of this operation to any arbitrary objects of the given universe will give rise to an object which is likewise an object of the given universe. For operation $\circ$ to be capable of guaranteeing the additivity of the quantities involved it must be *performable* in the class $(\hat{x})\,P_1x$, in other words the class $(\hat{x})\,P_1x$ must be *closed* with respect to operation $\circ$.[42] This means that if an object x is an element of the class $(\hat{x})\,P_1x$ and that if an object y is also an element of the class $(\hat{x})\,P_1x$, then the object which arises out of the combination of objects x and y (we denote the new object '$(x \circ y)$') must also be an element of the class $(\hat{x})\,P_1x$. Hence it holds that

$$(\forall x)\,(\forall y)\,\{[(x \in (\hat{x})\,P_1x).(y \in (\hat{x})\,P_1x)] \rightarrow (x \circ y) \in (\hat{x})\,P_1x\}.$$

The operation $\circ$ must be commutative in the class $(\hat{x})\,P_1x$, i.e.

$$(\forall x)\,(\forall y)\,C_{A_1}\,(x \circ y),\,(y \circ x).$$

The operation must be associative in the class $(\hat{x})\,P_1x$, i.e.

$$(\forall x)\,(\forall y)\,(\forall z)\,C_{A_1}\,[x \circ (y \circ z)],\,[(x \circ y) \circ z].$$

The operation $\circ$ must be *right-* or *left-invertible*.[43] We say that operation $\circ$ is right- or left-invertible if for each pair of elements x and y there exists an element z such that

$$(\forall x)\,(\forall y)\,(\exists z)\,C_{A_i}x,\,(y \circ z),$$
or
$$(\forall x)\,(\forall y)\,(\exists z)\,C_{A_i}(x \circ z),\,y,$$
respectively

There are a few more important properties of operation $\circ$ with respect to the predicates A_1 and C_{A_1}.

If it is possible to say of any objects that A_1x, y, then the same predicate A_1 can be related to a further pair of objects which arises out of the combination of objects x and y with a third object z, hence

$$(\forall x)\,(\forall y)\,(\forall z)\,[A_1x,\,y \rightarrow A_1\,(x \circ z),\,(y \circ z)].$$

If it is possible to say of any objects x and y that they coincide with respect to A_1, then objects $(x \circ z)$ and $(y \circ z)$ also coincide, i.e.

$$(\forall x)\,(\forall y)\,(\forall z)\,[C_{A_1}x,\,y \rightarrow C_{A_1}(x \circ z),\,(y \circ z)].$$

Since operation $\circ$ is right- or left-invertible, there must exist such an object z that

$$(\forall x)\,(\forall y)\,(\exists z)\,[A_1x,\,y \rightarrow C_{A_1}x,\,(y \circ z)],$$

or
$(\forall x)(\forall y)(\exists z)[A_1x, y \rightarrow C_{A_1}(x \circ z), y]$,
respectively

For these reasons it then holds that

$(\forall x)(\forall y)(\exists z)[C_{A_1}x, y \rightarrow A_1(x \circ z), y]$,
or
$(\forall x)(\forall y)(\exists z)[C_{A_1}x, y \rightarrow A_1x, (y \circ z)]$,
respectively

If a pair of objects x and y and another pair of objects u and v coincide with respect to A_1, then the two new objects which arise out of combination, $(x \circ u)$ and $(y \circ v)$, will also coincide, i.e.

$$(\forall x)(\forall y)(\forall u)(\forall v)[C_{A_1}x, y . C_{A_1}u, v \rightarrow C_{A_1}(x \circ u), (y \circ v)].$$

Of any quadruple of objects x, y, u, v, it further holds that

$$(\forall x)(\forall y)(\forall u)(\forall v)[C_{A_1}x, y . A_1u, v \rightarrow A_1(x \circ u), (y \circ v)],$$

and

$$(\forall x)(\forall y)(\forall u)(\forall v)[A_1x, y . A_1u, v \rightarrow A_1(x \circ u), (y \circ v)].$$

It is obvious that this system built up on universe $\mathscr{X}$ with the predicates P_1, A_1, C_{A_1} and the operation $\circ$ may also be built up axiomatically, in which case the formulae expressing the properties of these predicates and the operation $\circ$ will be the axioms and theorems of the system.

Thus it is possible to reconstruct a certain analogy between the system

$$\langle(\hat{x})\, P_1x, A_1, C_{A_1}, \circ\rangle,$$

and the system

$$\langle N, >^{44}, =, +\rangle,$$

where N is a set of real numbers. This analogy, however, does have limitations, which are rooted in methodological finitism, such as in the limited fineness of discrimination, the limited possibilities for establishing coincidence empirically, and so forth. For these reasons there is no way of carrying out in the first system the analogy of what corresponds in the second system to the principle of the density of real numbers, i.e. that for any pair of real numbers α and β ($\alpha \in N$, $\beta \in N$), where $\alpha > \beta$, there exists γ ($\gamma \in N$), such that $\alpha > \gamma > \beta$.

We have already pointed out that the quantification of properties assumes that if we can assign the quantity i to the relevant scale-factor m, then we are also justified in assigning the same quantity to an object x if x and m coincide with respect to the relevant comparative predicate. If P_1 is a qualitative predicate of the kind which allows of distinguishing various degrees of a given property (e.g. that of length, duration, heat or weight), then quantification may be treated as a function which assigns to P_1 a certain numerical value, or pair, 3-plet, ... n-ple of values, as the case may be, i.e.

$$f(P_1) = i,$$
$$f(P_1) = \langle i, j \rangle,$$
$$f(P_1) = \langle i, j, k \rangle \quad \text{etc.},$$

where $i, j, k, \ldots$ are real numbers. Since the function which assigns to P_1 certain numerical values always contains an operation with the respective scale-factors, any quantification assumes isomorphism between the quasi-series of scale-factors and a quasi-series of the given universe.[45] This also complicates the semantization of quantitative predicates. If we take as our basis the communication model which operates with blocks A, B and C, as described above, then we see that the semantization of quantitative predicates is quite a different matter from that of qualitative concepts. In determining the denotation of qualitative predicates we generally assign elements of block A. In the case of the semantization of quantitative predicates, however, this simple mode of assignment is just not sufficient. Account must always be taken of block B and the interconnections between blocks A and B. From this point of view, traditional semantics, which operates with the assignment of objects and classes of them (in block A) to expressions (elements of block C), cannot be deemed satisfactory. Hence, the semantization of quantitative predicates involves several aspects: (a) the assumption that the objects whose properties are to be quantified are objective and independent of the observer, measuring apparatus etc. (in other words the assumption that these objects belong to block A); (b) the assumption that block B is capable of registering and analyzing the properties of the objects that belong to block A; (c) the assumption that the analysis of the objects and their properties and relations by the devices available in block B can also be represented in a set of real numbers. By this the denotation of quantitative predicates is seen to be quite distinct from that of qualitative predicates since, in contrast to the simple assignment involved in the case of the latter, it contains at least three kinds of assignment in the sense of (a), (b) and (c).[46]

Quantitative predicates which belong to one and the same object or to all the objects of one and the same class are not as a rule independent of each other. The most primitive consideration given to the relations between weight, expressed in the relevant units of weight, and volume, expressed in the relevant units of volume, in a class of objects of the same substance will tell us that these relations are invariable for the entire class of objects. If we take the sign P for weight (taken as a quantitative predicate) expressed in the appropriate units, and the sign V for volume, expressed analogously, then it holds, for example, that

$$\frac{P_i}{V_i} = \frac{P_j}{V_j},$$

or, to put it generally, $P = kV$, where the constant k, which varies according to the substance involved, is called *density*. Since weight and volume can be quantified, it is also possible to quantify density on this basis.

It was similar considerations to these which led to the differentiation of fundamental and derived measuring: when a certain standard for $P_i/V_i = 1$ has been selected it is possible to obtain other numerical values by derivation from the relations given. The distinction between fundamental and derived measuring is of course conditioned historically in the sense that, for example, only weight and volume were quantifiable originally but on the basis of this it later became possible to quantify density. Another factor is that the quantities which express weight and volume are additive whereas those for density are not. However, the distinction of fundamental and derived measuring is only relative: if we use the communication model which operates with blocks A, B and C, then which of the properties we are able to determine as fundamental and which as derived depends on the equipment of block B. We can imagine, for example, apparatus including a densimeter and means of establishing volume. In the case of a liquid it is then a simple matter to establish weight by derivation, i.e. on the basis of density and volume. Thus the distinction between fundamental and derived measuring must be relativized with respect to the equipment of block B.

Taking as known blocks A and B and the limitations represented by the equipment of block B, we are in a position to determine (and quantify) a certain range of properties. In order to extend this range there are essentially two procedures open to us: either we must modify block B with respect to the constant A so as to be able to determine further properties, or we must avail ourselves of such *decision functions* as enable us to calculate additional properties on the basis of those which are determinable by the same block B without having to resort to additional blocks B′, B″ etc. What is used as decision functions are scientific laws: to calculate acceleration we use the formula $a = dv/dt$ where a is acceleration, v is velocity and t is time; to calculate tensile strength we use Hooke's law $\lambda = kp$, where λ is the lengthening of the bar and p the tension which is itself in direct proportion to the load on the bar, and in inverse proportion to the bar's cross-section.

It might seem from what has here been said about the communication model and its use in the interpretation of predicates and their relations in the language of science, that the basis is block B, i.e. the apparatus for registering properties, for the observation and measuring of properties, etc. In fact, however, the construction of measuring apparatus in the empirical practice of science is usually based on knowledge of decision functions, i.e. knowledge of the relevant scientific laws. Accordingly the communication model should be understood in the sense that any block B must be bound to a certain set of decision functions which effectually make possible two things: (a) they enable us to make decisions about elements of block A and their properties and relations on the basis of information accessible at the output of block B, (b) they enable us to make decisions about other properties and relations on the basis of known scientific laws which operate with the elements, properties and relations which can be established on the basis of (a).[47]

4. The Classification of Predicates: Empirical, Dispositional and Theoretical Predicates

(a) *Empirical and Dispositional Predicates*

One of the fundamental principles of empiricism has traditionally been asserted to be the requirement that in the language of empirical science the only proper (verbal) signs which are to be considered meaningful should be those which can be defined ostensively, or, as it was expressed traditionally, those which denote observable objects, properties or relations. The other signs were considered meaningful, and hence scientifically admissible, as long as their introduction into the language of science could be realized on the basis of signs which fill the above requirements. This formulation is in effect the analogue of the familiar thesis of the English empiricists that all ideas arise out of previous impressions or are composed of ideas which are themselves the direct product of previous impressions. The main difficulties in this conception of empiricism have not usually been connected with those signs where the connection with empirical procedures, in particular observation, measuring or the results of the individual's own activity, is immediate, but with those where it has to be reconstructed using certain appropriate procedures.

As far as predicates in the language of science were concerned, it was assumed that this conception could be used as a basis for distinguishing between what were called empirical (observational) predicates and predicates of other kinds. It is natural that this distinction assumed that we have available certain criteria by which to decide whether any given predicate is empirical (or observational) or not. In the earlier versions of empiricism, where the phenomenalistic view dominated, these criteria were often treated quite subjectivistically: if the observer is able, on the basis of his own observations and knowledge of the language of which the predicate P and the name of an object a are components, to come to a decision about a statement Pa, i.e. to accept either Pa or $\sim Pa$, then predicate P may be considered empirical.

It is clear that such a conception of empirical predicates cannot exclude the possibility of being interpreted purely subjectivistically. It also assumes the fulfilment of certain further conditions which themselves are beyond the traditional framework of empiricism: leaving aside the question of the requisite knowledge of a language including its syntax and semantics, let us consider the requirement that it be possible to make a definitive decision as to whether a is an element of the class $(\hat{x})\, Px$ or of the complement of this class. It is of course quite easy to imagine situations where no such decision for one reason or another is yet possible. Allowance should also be made for the fact that the boundaries between $(\hat{x})\, Px$ and $(\hat{x}) \sim Px$ are in some situations far from clear-cut (in such cases the predicate P is a vague concept), so that any decision is to a degree uncertain. Finally it should be added that the conception of empirical predicates described above explicitly counts with the possibility of no more than binary decision-making and with the corresponding logical rules, assumptions which in themselves are not of an empirical nature.

Modern science has long since ceased to count with only the observer's abilities. It places between the scientist-subject, who is always equipped with certain theoretical knowledge, and the studied object the most diverse measuring and experimental apparatus capable of refining to very high degrees the human ability to distinguish the various observable properties and assign them to various objects. It must further be emphasized that the greater part of these instruments and apparatus are themselves based on theoretical knowledge, the knowledge of certain laws, theories and so on. Telescopes, microscopes, spectrometers, instruments for measuring tension, resistance and the like, are not merely the extension, broadening or refinement of human sensory potential but above all the fruit of man's theorizing activity.

Thus it is clear that any relatively precise or objective analysis of what has been considered the empirical basis of the language of science, must objectivize and relativize the whole conception of this basis. This also applies to the sphere of predicates which, relative to certain conditions, we shall be considering as empirical. The objectivization and relativization of empirical predicates can be nicely demonstrated on the communication model of empirical activity, operating with the blocks A, B, and C and the interpretation presented above. The concept of 'empirical predicate' must always be relativized to the interconnections between A and B. Since the communication model assumes that block B, i.e. the 'observer channel' or more generally 'channel of science', always has certain finite and restricted characteristics which relate to the limitations of threshold values, the discrimination of fineness, capacities, memories, delays etc., the conception of the empirical predicate is also rendered finite.

With respect to a given block A there are two types of empirical predicates to be distinguished:

(a) direct empirical predicates, which allow of an immediate decision by the use of results furnished by block B (e.g. by mere observation it is possible to distinguish colours, weight can be distinguished by the use of scales, etc.);

(b) derived empirical predicates which allow of decision, with respect to the given block A, by the use of results furnished by block B together with the rules for decision-making which we have at our disposal in block C. It is a familiar fact that many empirically distinguishable properties can be established indirectly, by mediation, i.e. on the basis of the results of measuring and by the use of the appropriate decision rules. If we have a densimeter and apparatus for measuring the volume of liquids, it is easy to determine their weight.

Since it is possible to operate, with respect to one and the same given block A, with a quite large (though always finite) number of different blocks B, or with a global block representing the unification of all the separate blocks B_1, B_2, ..., B_n, together with quite a large number of decision rules, the range of what we acknowledge as empirical predicates may be extended accordingly. However, the delimitation of this sphere will always be relative, not only to block A, i.e. the given domain of objects, but also with respect to blocks B and C. Since blocks A and B implicitly also contain the results of theoretical activity, the delimitation of the range of empirical predicates is also indirectly relativized to the sphere of theoretical data included in blocks B and C.

From this point of view we might well treat as a failure any attempt at the construction of what has been called an empirical language (or analogous attempts to construct a physicalistic language, the so-called *Dingsprache* and so forth) which would contain as proper (verbal) signs 'observable terms', and in addition to these only such terms as are definable by the use of them. Advocates of these opinions, in particular adherents of the positivist movements and the earlier period of logical empiricism, justly found themselves under the fire of criticism. Russell,[48] for example, rebuked Carnap for his claim that for decisions as to 'observability' we never need more than a few observations. Similarly Putnam[49] objected that the concept of an 'observable predicate' cannot be taken as invariant even with respect to the given language of empirical science. From the point of view of the evolution of science it is a quite familiar fact that entities that were once considered to be 'unobservable' came in time to be registerable with accuracy thanks to improvements in measuring and experimental apparatus. One of the most significant objections to the conception of an empiricist language has been raised by Quine.[50] Quine speaks of the 'dogmas of empiricism'. One of these (two) dogmas is, according to Quine, the view that any meaningful expression, as long as it is synthetic, is equivalent to a certain logical construction composed of the elements of language which correspond to immediate sensory experience, in other words composed or constructed of 'observable' (verbal) signs. One of the weakest points of this conception of an empiricist language, still according to Quine, is the claim that the meaningfulness of any expression, i.e. the fact whether or not it is constructed of 'observable' signs, can be decided in isolation, irrespective of any other signs in the language of science or of any connections there may be with the latter.

The program of an empiricist language, i.e. the requirement that all meaningful expressions of the language of science be either 'observable' signs or signs constructed out of 'observable' signs,[51] has run into many difficulties ever since the outset. If we consider as 'observable' signs, and therefore also as atomic proper signs of the language of science those which permit of immediate ostensive definition, then the question arises as to how to construct molecular signs. [Note should be made of the viewpoint, which can scarcely be upheld in the practical structure of the language of science, that only 'observable' signs are admissible as atomic signs.] The simplest mode of constructing molecular signs on the basis of atomic signs seems to be explicit definition. For example, if P_1, P_2, P_3 are atomic signs, a new molecular sign can be defined as the conjunction of these primitive signs. Hence a square can be defined as a quadrangle with equal sides and right-angles. Notice that this way of introducing molecular signs into the language of science on the basis of atomic signs (which are assumed to be definable ostensively), in effect represents a kind of economization of means of expression. For in the case of explicit definition the definiendum which is supposed to be synonymous with the expressions which make up the definiens, is in fact an abbreviation for these expressions taken as a whole. A rather different kind of defining is represented by what are called contextual definitions. Whereas an explicit definition defines an expression in itself, the contextual definition represents a rule as to how to assign a synonymous sentence to the whole sentence in which the expression to be defined is used. We have already indicated one example of a contextual definition when dealing with the analysis of comparative predicates. The predicate 'to be harder than' is much better defined contextually: A substance x is harder than a substance y if and only if a crystal of x will produce a scratch on a crystal of y, the reverse not being possible.

It is fairly easy to show that if the language of empirical science were composed only of observable signs, and signs which are definable from them, it would be very hard-up. Moreover a language like that would not be capable of designating the very properties which are generally the most significant scientifically, i.e. properties which are manifest only under certain circumstances. It can be easily shown that the use of contextual definitions runs into certain difficulties with what are called 'dispositional predicates'. Dispositional predicates denote what we call the disposition of the object, i.e. its ability to react in a particular way to a certain kind of stimuli. Let us now look at two contextual definitions from this point of view:

$$Qx, y = df\ Px, y \,.\sim Py, x\,,$$
$$Q'x = df\ P'x \rightarrow P''x\,.$$

(For the sake of simplicity we might take in the first case the interpretation already suggested, i.e. Q is the comparative predicate 'is harder than', P is the two-place predicate 'scratches'. In the other case Q′ is the dispositional predicate 'is soluble', P′ the empirical predicate 'is immersed into a liquid', and P″ is the empirical predicate 'dissolves in a liquid'.)

It is not very difficult to demonstrate that the second contextual definition is not satisfactory for reasons connected with the properties of material implication. [In the first case the construction of the molecular predicate is based on conjunction.] It is a familiar fact that material implication makes it possible to derive anything at all from a false antecedent. Let us assume that x is an insoluble substance which has never been immersed in a liquid. Then considering the properties of material implication, we might assert that x is soluble, since a false antecedent in material implication implies anything. From this it is evident that at least as far as dispositional predicates are concerned the original programme of an empiricist language ran into serious difficulties. And yet such dispositional predicates as 'soluble', 'inflammable', 'conductive', 'resilient', 'firm', 'resistant to corrosion', 'transparent' etc., are of exceptional importance in the empirical and experimental sciences.

The first attempt to solve these difficulties, and hence the first concession from the original program of the buiding-up of an empiricist language, was Carnap's well-known work *Testability and Meaning*[52] which suggested replacing definitions, in the case of dispositional predicates being introduced, by what he calls reduction sentences. The latter remove the most serious flaw of the contextual definitions that operate with material implications, namely the possibility of stating, for instance, that a substance which has never been immersed in a liquid is soluble.

Carnap presented the two most important forms for the introduction of dispositional predicates (Q) with the aid of empirical predicates (P, P′, P″, P_1' etc.): (a) the bilateral reduction sentence, and (b) the reduction pair. In the simplest cases the bilateral reduction sentence takes the form

$$(\forall x)\,[\mathrm{P}'x \rightarrow (\mathrm{Q}x \equiv \mathrm{P}''x)]\,.$$

It is generally considered expedient to give also certain conditions such as that the process is realized in a certain interval of time, with a certain intensity which exceeds the liminal potential of discriminability etc. These conditions are accordingly given as an additional argument in empirical predicates, i.e.

$$(\forall x)\,[\mathrm{P}'x,\, k \rightarrow (\mathrm{Q}x \equiv \mathrm{P}''x,\, k)]\,,$$

where k are the required conditions. In the case of reduction pairs, equivalence is replaced by material implication, the negative eventuality also being considered, i.e. $\sim$ Q. The simplest form of a reduction pair is

$$(\forall x)\,[\mathrm{P}_1 x \rightarrow (\mathrm{P}_2 x \rightarrow \mathrm{Q}x)]\,,$$
$$(\forall x)\,[\mathrm{P}_3 x \rightarrow (\mathrm{P}_4 x \rightarrow \sim\mathrm{Q}x)]\,.$$

Reduction sentences are capable of expressing a disposition on condition that the necessary experimental conditions have been ensured, while with the use of a contextual definition with material implication objects have any disposition at all if the necessary experimental conditions have not been ensured. Reduction sentences thus express the realization of the relevant conditions, i.e. the effectuation of certain operations. For these reasons reduction sentences are also considered as the formalization of what Bridgman and the adherents to operationalism have called operational definition. It must of course be stressed that reduction sentences are not definitions in the proper sense of the word, nor can they replace definitions. It is after all necessary to respect what is sometimes described as the *openness* of concepts in the language of empirical science. The openness of dispositional predicates can be demonstrated by the following consideration: Let Q be a dispositional predicate introduced with the aid of (ostensively definable) operations expressed by the empirical predicates P_1' and P_1''. The respective bilateral reduction sentence has its barest form – leaving aside quantifiers and arguments:

$$P_1' \rightarrow (Q \equiv P_1'').$$

This same disposition may be established also by the use of other similar, related, or sometimes even quite dissimilar operations. This is the case in the establishment of, for example, resilience, conductivity etc. Hence we are usually in a position to refer to a certain class of reduction sentences, which, however, is not closed, namely

$$\begin{aligned} &P_1' \rightarrow (Q \equiv P_1''), \\ &P_2' \rightarrow (Q \equiv P_2''), \\ &\quad\vdots \\ &P_n' \rightarrow (Q \equiv P_n''). \end{aligned}$$

It will be apparent that reduction sentences express the connection between the disposition and the contingent connections between properties that can be ascertained empirically. The performance of a certain number of steps in which this contingent connection is verified does not, however, exclude the possibility of further steps leading towards the same disposition. If we treat reduction sentences as empirical interpretations of dispositional predicates, this means that we are able as a rule to assign more than one empirical interpretation to one and the same dispositional predicate.

This possibility of a number of different empirical interpretations of dispositional predicates is a radical contrast with the original principles of Bridgman's operationalism whereby one and the same operation defines one and the same concept, and different operations define different concepts. If this requirement were to be fulfilled it would be neccessary to have at one's disposal distinct concepts including distinct dispositional predicates for each distinct empirical context. The realization of Bridgman's requirement would quite understandably lead to the pointless multiplication of the vocabulary, and many distinct expressions would in fact have to designate a different disposition whenever the disposition were demonstrable in different empirical contexts.

The pointlessness of multiplying the vocabulary is also to be seen from the fact that one and the same disposition, although demonstrable in a variety of empirical contexts, can generally be explained on the basis of the same set of theoretical data. We say that a given object is soluble in water if, immersed in water, it dissolves. Solubility, however, is a property which is bound to a certain sphere of atomic-structure properties of substances, usually expressed by theoretical data operating with what we call theoretical predicates. For this reason some authors [53] go even further and place the requirement that when dispositional predicates are introduced, the respective theoretical data be also respected. Therefore it is also possible to draw conclusions about certain dispositions not only on the basis of operations that have already been carried out (expressed here in terms of the connections between the predicates P_1' and P_1'', P_2' and P_2'' and so on), but even before such operations have been carried out. For example, we consider as inflammable an object which, with a certain degree of probability, will ignite if we expose it in the process of various experiments to the diverse effects of high temperatures. However, we may also conclude that the given object is inflammable on the basis of other discoveries and theoretical knowledge, for example, knowledge of its chemical structure, the ability of the substances it contains to combine with oxygen, and so on. Having this knowledge at our disposal we will be willing to consider this object inflammable even if the classes $(\hat{x})\ P'x$ and $(\hat{x})\ P''x$ are empty. A disposition then requires us to expect (the expectation being justified either theoretically or by previous instances) that if the operation coming under class $(\hat{x})\ P'x$ takes place, there will be a response which comes under class $(\hat{x})\ P''x$.

Among the considerable amount of literature devoted to the relations between empirical and dispositional predicates we sometimes come across attempts – in contrast to the ordinary conceptions where dispositional predicates are reduced to empirical – to treat even empirical predicates as dispositional and so effectively reduce all empirical properties to dispositions. These attempts are based on the undisputed fact that there are generally certain prerequisites behind any differentiation of empirical properties, i.e. to be able to distinguish properties of colour we must not be colourblind, which is to say that we must have at our disposal the normal sensory equipment, suitable light conditions, etc. In the same spirit Goodman,[54] for example, emphasizes the dispositional character of all sensile qualities. He points out that if we see, for example, something green, it is merely that certain conditions have been fulfilled, since under other circumstances, e.g. in another light, we might see the same object as blue. From this it seems obvious that a dispositional character might be ascribed to practically all empirical predicates.

This criticism of empirical predicates and attempts at reduction in the opposite direction are of course aimed first and foremost against the original version of the empiricist language, which assumed a stable and unvarying basis for reduction based on the normal human sensory equipment and the human ability to discern individual properties. In this respect the criticism and the associated controversy about empirical predicates are not unfounded. If we take a look at the examples given of 'observable' properties we readily see that their discernibility is bound not only on the abilities of the human sensory equipment but also on certain conventions and semantic rules of the respective language. However, we have already indicated that the conception of empirical predicates must be relativized – on the one hand with respect to the links between blocks A and B, and on the other hand with respect to block C. Viewed in this way it is therefore expedient to distinguish between those predicates about which we can make decisions directly at the output of block B – bearing in mind, of course, the appropriate relativization – and those which express our empirically or theoretically founded expectation that the given object will react in a particular way if and when certain operations take place.

(b) *Theoretical Predicates*

Although the admission of reduction sentences meant a concession to the cost of the original empiricist program, and although it became essential to recognize what is called the openness of the concepts of the language of science, it has proved impossible to treat all the predicates of the language of empirical science in this way. All of the empirical sciences operate – in addition to empirical and dispositional predicates – with many other predicates which cannot be reduced to either empirical or dispositional predicates and which are generally described as theoretical predicates. These are sometimes also known as theoretical constructs, the expression 'construct' hinting at the constructive nature and constructive origin of this kind of predicate. Attention is drawn to the fact that for the precise formulation of the dependences which characterize the relations of empirical data it was necessary to construct (verbal) signs capable of designating these dependences with greater accuracy. Examples are often adduced from geometry and physics: in nature as such there are no perfect circles, squares etc. In order to obtain more precise information about the real formations which approximate to circles, squares, straight lines, planes and so forth, we are led to construct ideal objects which do not exist as such in nature but which do enable us to determine with greater accuracy the properties of the real formations. Likewise physics operates with 'mass point', 'ideal gas', 'an absolutely black body', 'solid body', 'ideal steam engine' and others, even though no such entities can be found in nature having exactly the properties which theory ascribes to them.[55]

Theoretical predicates cannot be introduced into the language of empirical science by the use of explicit or contextual definitions, nor by the aid of reduction sentences or strings of reduction sentences. In the effort to save the original programme for constructing an empiricist language, or the operationalist program as the case may, certain authors have pointed out the possibility, when theoretical predicates are being introduced, of admitting even such operations as do not have a physical character (as was the case with the dispositional predicates, i.e. 'to immerse in liquid', 'to dissolve', etc.) but are of a logical or mathematical nature. Considerations in this sense have usually operated with quantitative data and been based on the appropriate theoretical knowledge. Thus Bridgman, for instance, in his later writings admitted, in the endeavour to uphold the principles of the original program, what he calls 'paper-and-pencil' operations.[56] With like purpose others speak of mental operations, such as processes of abstraction, idealization and so forth. It is clear, however, that such attempts to save the original principles of the empiricist or operationalist program are, on the one hand, in fact in conflict with the original intentions, and on the other hand lead to the highly dubious confusion of physically realizable and hence ostensively definable operations with operations of a logical or mathematical nature.

There have also been attempts to eliminate theoretical predicates from the language of the empirical sciences in so far as they are reducible to predicates that are ostensively definable. However, it has been shown that these cannot claim to be general. In particular they are not able to capture the inductive connections between immediately verifiable statements having an empirical content and statements from the theoretical and deductive system.[57]

The most common way of introducing theoretical concepts into the language of science is what are called implicit (sometimes also know as postulatory or axiomatic) definitions. This method has long been applied in mathematics and the formal sciences generally. It enjoys a particularly well-tried tradition in geometry and number theory. The meaning of theoretical predicates is not determined explicitly here; it is given implicitly, i.e. by the use of postulates or axioms in which the theoretical predicates occur. For instance, in the case of the familiar postulates of geometry there is no call to assume beforehand that expressions such as 'point', 'straight line', 'plane', etc. denote concrete objects. The postulates of geometry fix the relations between these expressions and hence determine their meaning implicity. The procedure is analogous in mathematics and logic. For example, the predicate 'transitive' is determined in such a way that it denotes the property of the predicate P if it is true of any triple of objects in the given field of objects that

$$(\forall x)\,(\forall y)\,(\forall z)\,(\mathrm{P}x, y\,.\,\mathrm{P}y, z \rightarrow \mathrm{P}x, z).$$ [58]

It should be recalled that the major triumphs of the natural sciences in modern times begin whenever they succeed in collating their data into deductive-theoretical systems in which they then operate not only with empirical but also with theoretical predicates. Also the meaning of these predicates is then declared by implicit definitions. This is the case with the basic concepts of Newtonian dynamics, for instance. Furthermore, deductive-theoretical systems conceived in this way could also be constructed as systems of axioms which are implicit definitions of the respective sets of concepts. This way may be used to build up, for example, the axiomatic system of Newtonian dynamics in which some of the verbal signs would be of the character of fundamental theoretical predicates (e.g. 'force', 'mass', 'particle', 'time interval'), others having that of definable predicates (e.g. 'acceleration'). Certain physicists might justifiably object that this distinction could be changed. If we were to take, for instance, 'mass' and 'acceleration' as fundamental, it would be possible by Newton's second principle to define 'force'. Similar situations may also arise with certain other predicates. The selection of theoretical predicates as fundamental predicates of an entire deductive-theoretical system usually depends on the overall conception of the latter.

More serious problems arise if we are to use theoretical predicates which have been defined only implicitly, in an empirical context. It is quite natural, for example, that predicates that have been defined implicitly in the postulates of geometry, i.e. 'point', 'area', 'straight line' etc., have to be used to demonstrate genuine spatial relations. It is a question whether what is used are these or perhaps some other predicates. In the same way objections may be raised against the use in singular statements of those physical predicates – in so far as they have been defined only implicitly – which relate to the force of a given process, the mass of a certain object and suchlike. Adherents to the radical empiricist and operationalist programs must reject any such usage. They would point out that these are (verbal) signs which do not denote anything and hence are not meaningful empirically.

It is apparent that in so far as we consider theoretical (verbal) signs in isolation, it is difficult, indeed usually impossible to give them an empirical interpretation in any of the ordinary ways, i.e. using explicit or contextual definitions, reduction sentences or any other devices which operate with empirical predicates. It is necessary to respect the basic characteristic feature which is of the essence for the analysis of the meaning of theoretical predicates, namely that the implicit definitions by which the latter are introduced must be taken in their entirety. This means, for example, that we must take the whole of the postulates of geometry which, taken as a whole, fix the meaning of the theoretical predicates within the framework of the relevant deductive-theoretical system. From this point of view, if we are concerned with semantic analysis, it is scarcely even possible to think in terms of individual, isolated, theoretical predicates, but only in terms of their place and function within the relevant deductive-theoretical system.

The next stage, which on the one hand is a refinement of the analysis of theoretical predicates and on the other hand their relativization with respect to a given empirically apprehensible domain, is the introduction of what is described in the literature as 'interpretative sentences', 'correspondence rules',[59] 'coordinative definitions'[60] and the like. These represent a refinement of the procedures which enables some of the theoretical predicates to assign not a total but a partial interpretation. These are, for instance, the rules which make it possible to assign to real objects the properties dealt with by geometry, theoretical physics, etc., in an abstract form. Interpretative sentences represent a partial interpretation of theoretical predicates in the sense that they ensure the applicability of the deductive-theoretical system to a given sphere of objects. Applicability here means that on the basis of statements containing empirical or dispositional predicates and by the use of the respective deductive-theoretical system and suitable interpretational rules we are able to derive other statements containing empirical or dispositional predicates, that is we are in a position to determine other empirically discernible properties on the basis of the relevant theory, or at least to foresee such properties.[61]

Interpretational rules in the natural sciences are usually of the nature of statements which join together empirical and theoretical predicates. From the kinetic theory of gases, for example, we know that the average absolute temperature of a gas[62] is directly proportional to the kinetic energy of the molecules of which it is composed. If we assume that the predicate 'kinetic energy' is defined implicitly in the given theory, while the predicate 'absolute temperature' can be reproduced in terms of empirical data, then the sentence in question gives us a partial interpretation of the relevant theoretical predicate.

This emphasis on the fact that we are here dealing with a partial interpretation means that theoretical predicates, whose meaning must always be taken with respect to the relevant theory, i.e. the respective deductive-theoretical system, in its entirety, may have a variety of different interpretations in various empirical contexts. This circumstance is especially familiar in physics. It is generally pointed out that the structures of different real systems are similar so that these structures and the behaviour of the systems can be explained by the same deductive-theoretical system.

This is a fitting place to point out two distinct meanings of the concept of 'theory'.[63] In the formal sciences it is usually taken as meaning the deductive-theoretical system itself as given by the set of postulates or axioms which define implicitly the (verbal) signs used. This is the sense meant when we speak of group theory, set theory, matrix theory or automata theory. In this respect the concept 'theory' is understood purely analytically. The individual theoretical predicates are operated with in these deductive-theoretical systems without either the individual or possible empirical interpretations being stated, or respect being paid to the empirical nature of the universe to which a given theory may be related. This is the sense in which we operate with 'points', 'straight lines', 'planes' and so forth in geometry.

In contrast to this the concept 'theory' is understood much more widely in the empirical sciences. On the one hand it is the set of postulates which define the theoretical predicates implicitly and enable the respective deductive operations to be carried out, i.e. the deductive-theoretical system in the sense just stated, but in addition it is also a set of interpretational or correspondence rules. If we call the former T and the latter C, then theory in this second sense effectively means $T.C$ where it is taken as self-evident that $T.C$ makes a consistent system. Thus references to the kinetic theory, the theory of relativity, quantum theory and suchlike usually mean theory in the second sense.

We have already emphasized that interpretational rules represent a partial interpretation of theoretical predicates. This means that with respect to one and the same deductive-theoretical system there may be more than one set of interpretational rules, i.e. $C_1, C_2, \ldots, C_n$. It then depends on factors of a pragmatic nature whether we consider $\langle T, C_1 \rangle$, $\langle T, C_2 \rangle, \ldots, \langle T, C_n \rangle$ as a single theory, as a number of 'related' theories, or as different theories. This certainly does not eliminate the need to modify the original theory T, i.e. to move from T to T', in the case of a transition to a different theoretical domain.

The development of thinking in the natural sciences furnishes some interesting samples of how new interpretational rules have been added to the same deductive-theoretical system T much more often than new deductive-theoretical systems have been created. Not until an extra $\langle T, C_i \rangle$ proved to be inconsistent was it necessary to set about the much more difficult task of elaborating a new theory.

At times we encounter in the literature interpretational rules treated as if they were empirical definitions of theoretical predicates. While this accorded with the original empiricist and operationalist program, it has been shown to be an incorrect approach. If we were to abide strictly by the operationalist principles we should have to bring up a different theory for each empirically distinct domain of objects. Since different operations are capable of defining different concepts, in the operationalist view, we should have to operate with so many expressions for heat as we have different procedures for its measurement, which is an objection which has been raised a number of time against this view. It would similarly be necessary to assume that for each different set of interpretational rules we should have to have a different deductive-theoretical system operating with different theoretical predicates. This would then make it impossible to operate in the physical sciences with only one expression for energy or entropy and others. Thus it should be apparent that if the operationalist program were carried out consistently it would lead to the violation of Occam's razor and its semantic analogue whereby *nomina non sunt multiplicanda praeter necessitatem.*

The discovery that theoretical predicates cannot be introduced into the language of empirical science either by the use of explicit or contextual definitions or by the use of reduction sentences, as the case may be, means that it becomes impossible to explicate empirical predicates by the use of theoretical concepts.[64] Such approaches are usually connected with the explanation of the functions of the given deductive-theoretical system and the theoretical concepts of which it consists. Carnap, for instance, has pointed out that the properties of the various chemical substances which are empirically distinguishable and ostensively definable can be explained as the configuration of particles of a particular type, the typology of these particles operating with theoretical predicates. Likewise it is quite common to explain thermic phenomena by the use of energy phenomena, and so forth.

5. Similarity and Identification of Objects

(a) *The Semantic Background*

The identificational procedures that are applied in the empirical sciences may be of one of two kinds. Either it will be a matter of identifying extra-linguistic objects, i.e. the entities which we consider as elements of block A in the communication model, or it might be a question of identifying expressions of language or their component parts, which we conceive as elements of block C. In the former case the procedure of identification has the character of ontic decision, in the latter that of semantic decision.

The distinction between objects that are identifiable (for instance on account of their having all their properties in common or because no available criteria are capable of distinguishing them) and objects that are not identifiable may be of some importance in both practice and theory. It may be of significance with respect to the needs of classification in so far as it is our desire to gather into separate classes objects that can be identified. And there are practical tasks where we come across the requirement of identifying objects with given standards, for example in the case of quality checking of goods at the end of a production line. Procedures for identification in both scientific and practical tasks can be carried out only on the condition that we have available a finite number of distinguishing criteria, that we are capable of deciding as to identity in a finite time and with a finite collection of means for the purpose. In other words these and similar tasks require that we respect the demands of methodological finitism.

It is absolutely essential to keep apart the problem of the identification of expressions of language – this is usually associated with the problems surrounding synonymy – from tasks which have the character of ontic decision. Even here, of course, it is expedient to make decisions on the basis of a finite analysis of criteria. The conception of these criteria in modern logic and logical semantics continues in the tradition of Leibnizian philosophy and the criteria which Leibniz himself formulated. However, the problems concerning the identification of expressions of language (whether identity of simple or compound expressions), on the one hand, have sometimes – in the spirit of certain Leibnizian traditions – not been kept strictly separate from those relating to the identification of things or any other extralinguistic objects, on the other hand. That is to say that sometimes there has been a confusion of identificational tasks having the character of ontic and semantic decision. Although the two complexes are closely connected, in the main on the basis of relations of a semantic nature, there is some point in keeping them strictly apart.

In the case of the identification of expressions of language, Leibniz' formulation *Eadem sunt quorum unum potest substitui alteri, salva veritate*[65] is usually taken as a guide. This criterion for the identification of expressions of language finds the guarantee of identifiability in the interchangeability of the expressions without change in the truth value. It is not difficult to prove, as the entire evolution of logical semantics to date as represented by such names as Frege, Russell, Lewis, Carnap and Martin has indeed shown, that this criterion is very narrow. Its formulation is based on one very simple thought: for two expressions to be identical as to semantics, i.e. for them to be considered as expressions having the same meaning and hence being interchangeable, we must assume that together with this interchangeability *something is preserved.* Leibniz' original formulation, which is sometimes also known as the *salva veritate* test, assumes the preservation of the truth value. Since in logical semantics the truth value is considered to be a certain semantic characteristic of sentences, i.e. their denotation or what is called their 'extension', we can generalize the formulation, which might otherwise be described as the *extensional criterion* of the identifiability of expressions, as follows: expressions are identifiable (extensionally) which can be interchanged, denotation (extension) remaining the same. The nature of the expressions to be interchanged is naturally of great significance since different kinds of denotation are distinguished in the different kinds of expressions.[66] However, not even this formulation for the criterion of identifiability of expressions of language, which is based on their semantic characteristic, namely that they *denote something* in the extralinguistic world, and which requires that in the case that they are replaceable by another expression this semantic characteristic remain unchanged, has proved satisfactory since it too fails to answer to our intuitive conception of expressions of language. An intuitive conception must have stricter criteria for identifiability, which is usually understood as synonymity. The usual demand is for the conservation of another semantic characteristic which might be described as sense or intension. The differentiation of these two semantic characteristics, which was first given a relatively precise form by Frege, did lead to the solution of some, but not all, of the difficulties which arise with the use of the narrow, i.e. only extensional, criterion of identifiability. These difficulties have been described in the literature as the 'antinomy of synonymous names'.[67]

Attempts have been made at a more precise definition of criteria of identifiability for expressions of language. They have been based on the results of logical semantics and in one form or another they hark back to Frege's distinction of *Sinn* and *Bedeutung*. However, they have not proved capable of providing solutions which would eliminate all difficulties, such as those connected with complex expressions based on 'non-extensional contexts'.[68] These attempts have also failed to distinguish interchangeability in differents modes of the use of expressions, as has been shown by Quine in particular. Nevertheless they have pointed in certain main directions where to look for further refinements of the criteria of identity. As *identifiable* we consider expressions which have *certain common properties* of which the main ones are (a) *interchangeability* – what is understood by this must of course be defined with care, as also the contexts and modes of use[69] in which it is admissible, (b) the *preservation of certain semantic properties* of the expressions, where again the precise definition of the properties for the individual kinds of expressions is desirable.

Although we shall be endeavouring in the remainder of our account to respect this briefly indicated set of problems connected with the identification of expressions of language, the focus of our attention will be on another complex of problems, which is a special case of ontic decision, namely the problem of the identification of extralinguistic objects. We shall be adhering to the principles of methodological finitism in the sense we have outlined above. The conception of the identification of objects with which we shall be concerned is not to be confused with the traditional problem of identity in logic. In spite of our considering the procedures of identification in this conception as a special instance of ontic decision, we shall continue to lay emphasis also on certain pragmatic aspects. This means that besides the finite set of identificatory criteria we shall also be respecting the aims of identificational procedures as well as the demands placed on their precision where this is relevant. In terms of this conception *we shall not be speaking of identical objects, but of identifiable objects.*

In the identification of objects it is common to operate with the concepts of 'property' or 'class'. From this point of view we may confront the following formulations which are an expression of such an intuitive standpoint:

If x and y are two objects, then x is identifiable with y if and only if x

has all the properties which y has and if y has all the properties which x has.

Objects x and y are identifiable if and only if x and y have all properties in common.

Objects x and y are identifiable if and only if x has any of the properties that y has and vice versa.

Certain other alternatives to these formulations would also be possible, but they would be no more than different, more or less fitting expressions of the same principles. The formulations which we have given so far as expressing the intuitive viewpoint have all operated with the concept of 'property'. Since we might fairly assume that on the basis of the principle of comprehension there is a class to correspond to any particular property (which is not to say that such a class must necessarily be non-empty), we might make some other alternative formulations operating with the concept of 'class'. The assignment of a property to a given object, or the statement that this object is an element of a given class, always assumes that we are stating something about that object.

The conception of the identifiability of objects which operates with the concepts of 'property' or 'class' may meet with certain difficulties or objections. Some of the latter might be pronounced in terms of the extreme nominalist view. If we operate with the concepts 'property' or 'class' then our definition of the identifiability of objects may implicitly indicate the existence of properties or classes, which from the point of view of extreme nominalism is denied. Since identifiability itself is a property of at least two objects, this objection means that the very conception of identifiability itself is *eo ipso* suspended.

Another objection to the conception given here might be formulated from the standpoint of methodological finitism. In order to examine all the properties involved we must have at our disposal the means to be able to do so in a finite time using a finite set of the means. If the number of properties (and/or classes) is very high, or if it approaches infinity, this task may be more than can be coped with – from the viewpoint of methodological finitism.

If we limit ourselves to the set of properties which are denotable by one-place predicates and introduce the two-placed predicate 'is identifiable' (I), then

$$(\forall x)\,(\forall y)\,(\forall \mathrm{P})\,[\mathrm{I}x,\, y \equiv (\mathrm{P}x \equiv \mathrm{P}y)]\,.$$

Leaving aside certain logico-semantic problems connected with the quantification of predicates, we find that this schema of the identificational procedure has certain difficulties. For example, the possibility cannot be excluded that the objects have an infinity of properties or at least so many that we cannot handle them all with the means and delays available. No human being nor any technical apparatus is capable of distinguishing and ascertaining an infinite number of properties in a finite time and with finite experimental and linguistic means. It is scarcely conceivable that a human being, who never has more than finite 'capacities', 'delays' and 'memories' at his disposal, would ever be able to state about any object everything that it would be possible to state but for the limitations of methodological finitism. If we were to use our schema of the communication model, we should have to use, with regard to the block A whose objects we wish to identify, an infinitely perfect block B or an infinite class of blocks B with an unlimited ability to discriminate. From all this it must be apparent that we are bound to respect the limitations imposed by methodological finitism.[70]

Our schema of the procedure of identification, which is based on the assumption that it is possible to identify any two objects which have all their properties in common, or objects of which it is true that anything that can be stated about one of them can also be stated about the other, is not quite correct as it stands. If we also treat a property as a relation, i.e. as the property of a pair, 3-plet, ..., n-ple of objects, then the schema represents a simplification, being, as it is, confined to one-placed predicates.

Assuming that the class $\mathscr{P}$ of all predicates can be broken down into the class of one-placed predicates $\mathscr{P}^{(1)}$ $(\mathrm{P}^{(1)} \in \mathscr{P}^{(1)})$, that of two-placed predicates $\mathscr{P}^{(2)}$ $(\mathrm{P}^{(2)} \in \mathscr{P}^{(2)})$, up to that of n-placed predicates $\mathscr{P}^{(n)}$ $(\mathrm{P}^{(n)} \in \in \mathscr{P}^{(n)})$, the schema of the procedure of identification might be expressed more adequately as

$$(\forall x)(\forall y)(\forall \mathrm{P}^{(1)})(\forall \mathrm{P}^{(2)})\ldots(\forall \mathrm{P}^{(n)})\,[\mathrm{I}x, y \equiv (\mathrm{P}^{(1)}x = \mathrm{P}^{(1)}y) \\ .(\mathrm{P}^{(2)}x, u \equiv \mathrm{P}^{(2)}y, u).(\mathrm{P}^{(3)}x, u, v \equiv \mathrm{P}^{(3)}y, u, v)\,. \\ .(\mathrm{P}^{(4)}x, u, v, z \equiv \mathrm{P}^{(4)}y, u, v, z)\ldots(\mathrm{P}^{(n)}x, u, v, z, \ldots \\ \equiv \mathrm{P}^{(n)}y, u, v, z, \ldots)]\,.$$

This notation, which gives a much more correct expression of what might be demanded of the identificational procedure, indicates at the same time the difficulties with which we find ourselves faced if we fail to respect the principles of methodological finitism, with even greater clarity. Moreover, this kind of expression lays bare the risk of the uneconomical expansion of the vocabulary (and merely on the basis of combinatorics, for example) and so too the risk that the principle we have mentioned *("nomina non sunt multiplicanda praeter necessitatem")* might not be respected. This applies in particular to the necessity of constructing predicates with a high number of arguments, which, however, as we have already indicated, just does not correspond to the practice of the empirical and experimental sciences.

When we come to real tasks in science where there is a need for the identification of objects, it is essential to pay due respect to finitistic principles. This means, for instance, taking into account the possible and attainable degree of precision in measuring, the attainable degree of refinement in making distinctions, as well as other limitations associated with the properties of the measuring and experimental apparatus which is or may be available, the limitations of the time allowed for the solution of the tasks set, and so on and so on.

In conceiving criteria from the point of view of the principles of methodological finitism in the empirical sciences, the following should be taken into account:

(a) The potentials of block B, which we must have available in any task based on the identification of objects, i.e. its discriminatory ability, the limitations of the latter, its 'capacities', 'memories' and 'delays', are always finite. (This concerns not only the relevant parameters of the measuring and experimental apparatus but also the equivalent characteristics of man.)

(b) The goals of a given task, or the demands for precision, the reliability and quality of the data connected with these goals and important to the attainment of the goals, must also be taken into account.

(c) Due account must be taken of the varying degree of relevance of the individual criteria relative to the respective goals. (In this respect the actual ordering of the finite set of criteria, their placement in order of relevance, the differentiation of important and trivial criteria, etc., is very significant.)

(d) Finally account must be taken of the fact that the criteria need not be independent of each other and that certain dependences may subsist between them. Where it may be assumed that the criteria are not statistically independent of each other, the knowledge of their interdependences may well be of service in helping us to reduce the number of criteria needed.

(b) *The Finitistic Conception of Identification*

The simplest version of the identificational procedure is finding out that two objects do have all their properties in common. Disregarding the principles of methodological finitism this version of the identification of objects can be expressed using the schema we have given. The latter, however, requires that we work with quantifiers tied to expressions of a type higher than zero, which might evoke certain doubts and difficulties.[71] In order to avoid the latter and at the same time respect the way of thinking which is common in the empirical sciences, we shall be avoiding the quantification of expressions of the higher type, taking as a basis languages of the first order.

The simplest procedure is in a situation where we can confine ourselves to a finite class of predicates. Taking $\mathscr{P}$ as that finite class ($\mathscr{P} = \{P_1, P_2, \ldots, P_n\}$), and confining ourselves for the sake of simplicity to only one-placed predicates, we must investigate one by one all the properties that are distinguishable in the given universe and so consider all the elements of the class $\mathscr{P}$. If two objects are identifiable then we must demonstrate that if each individual property appertains to the first object, it appertains to the second object also, and vice versa, hence:

$$(\forall x)(\forall y)[Ix, y \equiv (P_1x \equiv P_1y).(P_2x \equiv P_2y)\ldots(P_nx \equiv P_ny)].$$

This formulation is only a slight modification of the schema given above, the difference being that we here respect the finite class of common properties.

In this connection the question might well arise as to what will happen if in the process of cognition we discover a new, hitherto unknown property, for example P_{n+1}. If we can add $(P_{n+1}x \equiv P_{n+1}y)$ to the sequence of conjunctive members, identifiability is upheld; if not, there is a breakdown in what is identifiable. Since cognition in the empirical and experimental sciences makes allowance for the fact that both in classes of objects and in classes of properties (including relations) which can be distinguished in these objects there will be cases of the discovery not only of new objects but also of new properties, that is allowance is always made for improvements in the power to distinguish, it is quite natural that a breakdown in what is identifiable has its place in the evolution of cognition. On the other hand it is also possible that objects which have been considered as different, on the grounds, for example, that they appeared different in different situations and so were given various names, will be shown to be identifiable after all.

An identificational procedure based on a finite class of properties which are represented by one-placed predicates, does not correspond to the usual practical procedures in identification. In *practical* procedures we usually identify objects which we are not capable of distinguishing in any respect. This is in effect the burden of Leibniz' principle of *identificatio indiscernibilium.* This principle is based on the identification of objects which in the light of *all the criteria of differentiation* remain indistinguishable to us. There are, moreover, as we shall proceed to show, two ways of proceeding: (1) either by comparing the objects which we are to identify directly with one another, or (2) by using for the purposes of comparison other objects which we treat as standards and whose place on the appropriate scale is known exactly.

Before proceeding to investigate the two versions of identification we should note a number of methodological aspects of the method of identifying the indistinguishable. The basis here is the view that the objects of science, the events of the object world do differ from each other. In order to be able to treat two or more objects as identical we must demonstrate that the objects do not differ from each other, that we cannot distinguish them – not even for all the criteria of differentiation we have available. (We assume, of course, that the class of accessible criteria of differentiation is finite.)

The role of criteria of differentiation may be taken over by the relations which enable us to order the objects of a given universe. In other words the role may be taken on by the relations denoted by comparative predicates. Comparative predicates make up a sub-set of two-placed predicates ($\mathscr{A} \subset \mathscr{P}^{(2)}$). As we have said previously these are irreflexive, asymmetrical, transitive and C-connected two-placed predicates, where the predicate C_{A_i} denotes relative coincidence in respect of the comparative predicate A_i. Since the class of criteria of differentiation must be finite, it is desirable that $\mathscr{A}$ ($A_i \in \mathscr{A}$) is also finite.

The simplest method of identifying the indistinguishable is the direct comparison with one another of the objects which are to be identified. If it can be shown that with regard to all the given criteria of differentiation two objects cannot be distinguished, then we are entitled to judge that in respect of the given class of criteria of differentiation ($\mathscr{A} = \{A_1, A_2, \ldots, A_n\}$) the given two objects are identifiable, i.e.

$$\begin{aligned}(\forall x)\,(\forall y)\,[&\sim A_1 x, y \,.\, \sim A_1 y, x \,.\\ &.\, \sim A_2 x, y \,.\, \sim A_2 y, x \,.\\ &\ldots\\ &.\, \sim A_n x, y \,.\, \sim A_n y, x \rightarrow \mathrm{I}x, y] \,.\end{aligned}$$

It is clear that we may use for the identification of a pair of objects (in the sense and with the limitations which we have indicated) the conjunctive linking of all relative coincidences as well. This means that $\mathrm{I}x, y$ in effect represents the conjunction

$$C_{A_1} x, y \,.\, C_{A_2} x, y \,.\, \ldots \,.\, C_{A_n} x, y \,.$$

In practice there are only a limited number of cases of the most elementary tasks where we identify objects by direct comparison one with the other. It is more usual to use for comparison further objects, which are here described as *scale factors*.[72] And it is further assumed that a certain criterion of differentiation corresponds to each particular scale factor. If $\mathscr{M}$ ($\mathscr{M} = \{m_1, m_2, \ldots, m_n\}$) is the class of all scale factors we shall assume a one-to-one assignment of the elements of class $\mathscr{M}$ to the elements of class $\mathscr{A}$. This will be expressed by like indices.

The class of scale factors should be treated as an appropriately selected subset of objects of the given universe. This means then that $\mathscr{M} \subset \mathscr{X}$. If then we conceive the method of the identification of the indistinguishable by means of the system

$$\langle \mathscr{X}, \mathscr{A}, \mathscr{M} \rangle$$

we can express the identificational procedure by the following schema:

$$\begin{aligned}(\forall x)\,(\forall y)\,[&\sim A_1 x, m_1 \,.\, \sim A_1 m_1, x \,.\, \sim A_1 y, m_1 \,.\, \sim A_1 m_1, y \,.\\ &.\, \sim A_2 x, m_2 \,.\, \sim A_2 m_2, x \,.\, \sim A_2 y, m_2 \,.\, \sim A_2 m_2, y \,.\\ &\ldots\\ &.\, \sim A_n x, m_n \,.\, \sim A_n m_n, x \,.\, \sim A_n y, m_n \,.\, \sim A_n m_n, y \rightarrow \mathrm{I}x, y] \,.\end{aligned}$$

This also means that – appealing to the possibility of establishing all the relative coincidences – Ix, y represents the conjunction

$$C_{A_1}x, m_1 . C_{A_1}y, m_1 .$$
$$. C_{A_2}x, m_1 . C_{A_2}y, m_1 .$$
$$....$$
$$. C_{A_n}x, m_n . C_{A_n}y, m_n .$$

Thus in both of the versions outlined the procedure of identification breaks up into a finite number of steps, the actual number of which depending on the number of criteria of differentiation. These steps can be ordered – and in many instances it is indeed very profitable to do so – in such a way that we respect at the same time the relevance of the criteria. The latter is of course always relative with respect to the goals of the given task or class of tasks and to the demands made on the quality of its, or their, solution.

Another fact which must be given due attention in the ordering of the steps is that the individual elements of class $\mathscr{A}$ (and so too those of class $\mathscr{P}$) may, in exceptional cases, be considered as the designation of mutually independent properties of the objects. If a certain dependence is known, and if it can be expressed exactly, it may make possible a reduction in the number of necessary steps.

(c) *Similarity of Objects*

Intuitively we may treat similarity as 'weaker' than identifiability. Where any objects which have all their properties in common are identifiable, we might treat as similar any objects that have at least some properties in common. Reverting to the above-mentioned method of identifying the indistinguishable, we might say that if objects that are indistinguishable with respect to all the criteria of differentiation are identifiable, then objects that are indistinguishable with respect to at least some criteria are similar.

These are of course only very rough descriptions of the concept of 'similarity'. Their inadequacy can be seen from the following questions: What does 'some properties' mean? On what does the choice of these properties depend? Is at least one property sufficient to ensure what we want?

Let us take a look first at the last of these questions. If $\mathscr{P}_a$ is the class of all the properties of object *a* and $\mathscr{P}_b$ the class of all the properties of object *b*, the formulation of the following criteria of similarity offers itself: Two objects are similar if and only if there exists a property P_i such that $(P_i \in \mathscr{P}_a).(P_i \in \mathscr{P}_b)$. In other words the solution is offered whereby two objects *a* and *b* are similar if and only if the intersection of $\mathscr{P}_a$ and $\mathscr{P}_b$ is a non-empty set. However, the view that two objects are similar if they have at least one common property cannot be considered at all satisfactory for at least two reasons:

(a) A predicate, that is the name of a certain property, can always be found whereby the property pertains to both objects. If, for instance, one object has a certain property while the other object does not, it is always possible to construct a predicate capable of designating both properties. Moreover it is obvious that at the very least the objects share the property of being elements of the same given universe.

(b) Intuitively we are generally willing to admit similarity not on the grounds that the objects are similar in just 'anything' but that they are similar in 'something', this something, for one reason or another, being of interest to us. Therefore we assume that a certain predicate P_i is given, or a certain class of predicates $\mathscr{P}$ ($\mathscr{P} = \{P_1, P_2, ..., P_n\}$), such that we can verify to a sufficient degree that $P_i x \equiv P_i y$ or $(P_1 x \equiv P_1 y).(P_2 x \equiv P_2 y)....$ $.(P_n x \equiv P_n y)$, respectively.

Two essential problems emerge from these considerations:

(α) the problem of the *selection of significant properties*, the term significant always assuming a certain relativization with respect to particular goals, demands and in general to everything that we mentioned in connection with the requirements characteristic for methodological finitism;

(β) the problem of the *degree of similarity*. On the understanding that it is possible to differentiate between more and less significant properties it is also possible to differentiate between pairs of objects that are more or less similar. Thus there is no avoiding the question of whether and how it is possible to measure and express the differences in the degree of similarity.

Although this conception seems to satisfy intuition, it can be shown that the number of predicates is no criterion for the measure of similarity. A proof of this has been carried out by Watanabe.[73] The core of his thinking is as follows: Taking as the starting-point for the solution of the problem of similarity the system

$$\langle \mathscr{X}, \mathscr{P} \rangle,$$

it is possible, if $\mathscr{P}$ is the class of all atomic predicates ($P_i \in \mathscr{P}$), to arrive at the class of molecular predicates $\bar{\mathscr{P}}$($\bar{P}_i \in \bar{\mathscr{P}}$) by means of logical combinations of the elements of class $\mathscr{P}$. It then holds that the number of predicates $\bar{P}_i$ which pertain to the pair of different objects x and y is a constant number which is independent of the choice of these objects. For this reason Watanabe confines himself to a class of predicates, which comprises a certain sub-class of $\bar{\mathscr{P}}$, which are, as he says, of greater importance. Watanabe's work, which is based on some of the means of the mathematical theory of information, is of course confined to one-placed predicates, possible connections between which are left out of consideration.

The assessment of the relevance of the properties or relations taken as criteria of differentiation depends above all on the nature of the given task, the goals associated with it and the claims laid on the quality of its solution, i.e. on circumstances of a pragmatic nature. Once this relevance is known – it being naturally not absolute but relativized to the given circumstances and possibly also to others – there are a number of ways of proceeding further.

(a) If it is possible to set up an order of relevance for the individual properties or criteria of differentiation, i.e. to order the class $\mathscr{P}$ and possibly also $\mathscr{A}$, in such a way as to begin the enumeration of their elements from the more important properties (or criteria of differentiation, as the case may be), it will be possible to pass on from the qualitative predicate 'similar' to the comparative predicate 'more similar'. One pair of objects is more similar than another pair if it can be found that they coincide in more relevant properties, or if they coincide with respect to criteria of differentiation which are more relevant. It is quite evident that this kind of comparison of the degree of similarity is only very limited. The difficulties arise as soon as we discover agreement in a fairly high number of properties or coincidence with respect to a fairly high number of criteria of differentiation where these are of varying relevance. Even if we are in a position to set up an order of relevance for the properties or criteria of differentiation, we must always count with the possibility that some objects will be similar in some properties, or coincide with respect to some criteria, that are at the beginning of the order of relevance and again in others that come towards the end of the order. This leads to the question whether dissimilarity in a major property can be balanced out by a proportionately larger number of similarities in less important properties.[74]

(b) For these reasons it would be to advantage, if of course it is at all possible, to assign to the individual properties or criteria of differentiation a certain gravity which would be measurable quantitatively. In order for the gravity of a greater number of less important criteria to be compared with the gravity of a smaller number of more important criteria, gravity would have to be additive. Taking w_i as a gravity of this kind assignable to the property designated P_i or the criterion designated A_i, it will be possible to operate with properties or with criteria of differentiation having different gravities. We will then be in a position to set up a certain similarity boundary: we will consider as similar all pairs of objects which are distinguishable according to n properties or criteria of differentiation where the sum of the gravities assignable to all the properties which the objects have in common, or all the criteria of differentiation with respect to which the objects coincide, does not exceed a certain permitted value $\mathscr{E}$. i.e.

$$\sum_{i=1}^{n} w_i \geqslant \mathscr{E} .$$

In a similar way it would be possible to distinguish between pairs of objects which would be more similar or less similar.

(c) A more detailed analysis of the properties or criteria of differentiation on which the scientific procedures of identification or establishing similarity are based will make it apparent that properties usually form interconnected 'families' ordered by certain relations. Objects which are bound by any such family of properties may be described as *similarity classes*.

One such 'family' of properties is made up of capacity, pressure and temperature, namely in the analysis of gases. The ordering relation for this 'family' is the Boyle-Marriotte equation. In classical mechanics the parameters of position in time and space, mass and force are ordered analogously. Taking this view, all scientific laws can be envisaged as ordering relations for 'families' of properties. In other words, in identificational procedures or in those for establishing similarity there is something to be gained not only from respecting the fact that the respective properties or criteria of differentiation are of varying relevance, but also that they are not independent. The contingency of the properties and criteria of differentiation on which our procedures of identification and the establishment of similarity are based can be expressed in scientific laws or be inferred from them.

This discovery justifies us in making the following conclusion: identificational procedures or those involved in establishing similarity of objects in science cannot be treated as purely empirical procedures. The very fact that it is of use to take account of the relevance of the various properties or criteria of differentiation on which we base our decisions, and in particular the fact that it is essential to respect the contingency of these properties and criteria demands that we take into account theoretical knowledge, especially that of the respective scientific laws. This means that any relatively reliable identification of objects or processes of establishing their similarity must also pay due attention to theoretical aspects of scientific activity, which goes to show that the rigid division into empirical and theoretical approaches in science is unfounded. So the successful solution of these procedures depends not only on our empirical and experimental equipment but primarily on the theoretical equipment with which we are able to solve them.

(d) *The Problems of Reduction in Identification Procedures and in Establishing Similarity*

Methodological finitism assumes that the set of means used in the solution of a given task is respected as finite. Referring again to the communication model, this means the finite and limited potential of block B including the theoretical equipment applied. In scientific activity and in practice, problems frequently arise that are connected with other limitations of block B, for example that of the time available for solving a task, that of its capacity, and so on. In such a case it proves necessary to carry out the respective task with a limited number of means, i.e. with respect only to a limited number of discernible properties, criteria of differentiation, etc. The necessary reduction must then be made in such a way as to preserve the quality of the solution of the task or at least to ensure that any difference in the quality of the solution does not exceed a certain tolerance. In many cases the necessity for reduction is imposed by the simple regard for the limited properties of block B. This may relate, for instance, to the limitations involved in the operational potential and memory of a computer, or to the limitations of the measuring apparatus, the limited time available for the task to be carried out, such as in diagnosis, etc.

We have so far been thinking in terms of identification procedures and procedures for establishing similarity in the system

$$\langle \mathscr{X}, \mathscr{P}, \mathscr{A}, \mathscr{M} \rangle,$$

where $\mathscr{A} \subset \mathscr{P}$ and $\mathscr{M} \subset \mathscr{X}$.

Reduction may relate not only to the set of properties, but also to the set of criteria of differentiation, as well as the set of scale factors on the basis of which we make our decisions. This means, for example, a shift from class $\mathscr{P}$ to class $\mathscr{P}'$ (where $\mathscr{P}' \subset \mathscr{P}$), from class $\mathscr{A}$ to class $\mathscr{A}'$ (where $\mathscr{A}' \subset \mathscr{A}$), etc. In practice this means leaving out certain (usually less relevant) elements of the given classes in such a way that the demands made on the quality of the solution of the given task will still be satisfied, or that at worst they will be reduced by a degree which we are willing to tolerate.

One of the simplest ways of reducing the classes is the situation where we can rely on the verified interdependence of the elements of the classes. The interdependence can be expressed by means of the functions $f_1, f_2, \ldots, f_n$, which may be interpreted broadly enough for their elements to include what are designated in the literature as 'meaning postulates'.[75] Meaning postulates are expressions in which the semantically significant relations and properties of the proper signs of a given language are fixed.[76] Meaning postulates as understood in the broadest sense of the expression may include any semantically significant relations between the elements of the classes which go to make up the system $\langle \mathscr{X}, \mathscr{P}, \mathscr{A}, \mathscr{M} \rangle$. Also, what we call scientific laws and all implicit definitions may also be understood as meaning postulates. Taking $\mathscr{F}$ ($\mathscr{F} = \{f_1, f_2, \ldots, f_n\}$) as the class of all known meaning postulates, it is expedient to proceed from the system $\langle \mathscr{X}, \mathscr{P}, \mathscr{A}, \mathscr{M} \rangle$ to the system $\langle \mathscr{X}, \mathscr{P}, \mathscr{A}, \mathscr{M}, \mathscr{F} \rangle$. This expansion of the system will then enable us to reduce the number of steps necessary to carry out identification procedures or procedures for establishing similarity.

Let us take a simple example of how elements of class $\mathcal{F}$ ensure a certain reduction of the elements of the other classes. If $\mathcal{X}$ is the class of objects made of one and the same material and if we are making a quality check of these objects, say at the end of the production line, we may operate with two criteria designated A_1 ('to be heavier') and A_2 ('to have a greater volume'). Also it holds that

$$(\forall x)(\forall y)(A_1 x, y \rightarrow A_2 x, y),$$

and simultaneously

$$(\forall x)(\forall y)(A_2 x, y \rightarrow A_1 x, y),$$

so that we may consider A_1 and A_2 as equivalent. It is obvious that in any such situation it is quite possible to operate with only one of the criteria of differentiation, the other being simply omitted. Accordingly in an identification procedure it is possible to leave out the operation which is to establish coincidence with respect to one of two criteria of differentiation.

The assumption that the elements of the classes which make up the system $\langle \mathcal{X}, \mathcal{P}, \mathcal{A}, \mathcal{M} \rangle$ are interdependent is of considerable significance in minimizing the number of steps which must be made in various procedures. If the interdependence of the criteria of differentiation A_1 and A_2 mentioned above holds true, and if we are making an identification or investigating similarity, we will arrive at just the same results if we leave one of the criteria of differentiation out.[77] We can think in these terms not only about properties and criteria of differentiation, but also about the reduction of the elements of class $\mathcal{F}$. There are numerous familiar cases in the natural sciences where one law involves another as a special instance of it. For example, the laws of relativist physics also include the laws of classical mechanics, as follows from the Lorentz transformation of field and the relation between velocity and the speed of light. In cases where velocity is negligible in comparison with the speed of light there is no need – in practical tasks and applications – to take account of the Lorentz transformations, that is a reduction is permitted without the quality of the solution being practically impaired at all.

Thus reduction leading to the omission of some elements, as we have tried to justify, is not associated with any impoverishment or limitation of the variety of what we are able to ascertain about a given sphere of objects. Moreover, particularly in the case of reduction of empirically discernible properties and criteria of differentiation and the exacting operations of establishing them – reduction which is justifiable on the basis of a deeper knowledge of the interdependence of these properties – far from being an impoverishment of our scientific image of the given sphere of objects, is more an improvement of it. Thus the aim of reduction as we have presented it is not merely the simplification of procedures, the reduction of the number of steps involved in them, or the impoverishment of our image of the given sphere of objects, but on the contrary the economization of the solution of scientific and practical exercises and the minimization of the number of steps needed for the solution.

So far we have been dealing with reduction procedures based on *knowledge of the interpendence* of the properties and criteria of differentiation with which we operate. The advantages of such procedures are beyond question, they can be given precise justification and can be trusted to be highly reliable. However, this interdependence must be constant and invariable for the entire sphere of objects. Attention is generally drawn to the fact that it is an interdependence of a deterministic type, though in fact, as is indeed well known, this is only partially the case in the empirical sciences. Many dependences are of the stochastic type; they may be variable, so that calculations based on them may be less reliable. Examples of such situations are the dependences of the symptoms ascertainable in medical diagnosis, those of the individual data in biological systems and even more so in social systems. Even in these cases it is often essential to reduce the class of the parameters on the basis of which we make decisions. If the actual dependences are not known, reduction is still possible if we are able to compare the quality of the solution[78] before and after the reduction is made. Let us imagine that we are solving a given task using three criteria denoted by A_1, A_2 and A_3. If we obtain the same quality using only the pair of criteria A_1 and A_2, or a quality which only differs from the original one in a tolerated degree, then we may make a reduction by omitting A_3 without actually knowing the exact dependences between the three criteria. This pragmatically substantiated procedure may also stimulate a theoretical analysis which might lead to the discovery of the dependences between the criteria.[79]

We might adopt similar viewpoints to those from which we have been looking at reduction procedures, for the purpose of assessing the reverse procedure, namely constitution. By constitution we mean above all that the classes which make up the system $\langle \mathscr{X}, \mathscr{P}, \mathscr{A}, \mathscr{M}, \mathscr{F} \rangle$ are enlarged by the inclusion of new elements. The history of science tells us of many cases where the use of known properties and criteria of differentiation led to no more than a rough distinction of objects, which satisfied neither the nature of the sphere of objects being investigated nor the demands placed on the quality of the solutions to the problems which arose out of that sphere. The need for constitution sometimes arose out of the need to improve the previous unsatisfactory quality of the solutions of the tasks. The need also presented itself whenever a given sphere of objects was extended. This was the case in physics, for instance, with the shift to the study of the micro-world. The objects of the latter could no longer be distinguished merely on the basis of the properties and criteria with which classical physics operated; new properties had to be introduced.

The problems involved in the reduction and constitution of elements of the various classes do not, as we have shown, relate merely to the pragmatic aspects of scientific procedures (moreover they need not be associated with identification alone but with many other procedures), but may also be connected with the theoretical aspects, especially if they have their source in the various elements of class $\mathscr{F}$ or if they lead to the necessity of making some change in the latter.

NOTES

[1] Quine (1960; § 19).

[2] The best testimony to this is in the doubts which arise if the differentiation between proper names and descriptions has to be marked by graphic means, e.g. by the use of an initial capital letter.

[3] Ryle (1957).

[4] Carnap, in his polemic with Ryle (Carnap, 1956a), spoke in this connection of the acceptance or rejection of a certain linguistic system. The problem of general and abstract entities, the problem of the existence of these entities, is solved by Carnap on the basis of the acceptance of a certain linguistic system and on the basis of the strict segregation of what he calls internal and external questions. It can, however, be shown that in science it is not usually a matter of the suitability or otherwise of this or that linguistic system, but a matter of fixing a certain universe which, with respect to the statements of science, is of an objective nature. It holds, moreover, that any differentiation between external questions, i.e. questions as to the existence and acceptance of the whole linguistic system, and internal questions, i.e. those concerning the existence and acceptance of entities within the framework of a certain system, is merely relative, since any system of entities may be broken up into

subsystems or, on the contrary, may be treated as a component in a higher system. For a more detailed criticism of Carnap's arguments see Tondl (1966b).

[5] This is sometimes referred to as the accepted ontology for a given system of means of expression.

[6] Quine speaks in this connection of ontic commitments to which we are bound by the use of a particular kind of variables, and thus he also speaks of accepting an ontology. See especially Quine (1943).

[7] Naturally enough, this rule has its exceptions. In many cases graphic or phonetic adaptations of names are desirable, for example phonetic transcriptions into some languages. Essentially, however, these are not translations but merely different modes of graphic or phonetic adaptation. A slightly different case is that of geographical names, especially those whose origin is clearly to be sought in descriptions. In such cases we have a dual function of these words – that as names and that as descriptions. The Czech designation 'Smrčiny' for a certain range of mountains is necessarily translated into German by the name 'Fichtelgebirge'. The share held by the two functions of the given words may of course vary from situation to situation.

[8] Another indication of the difficulties connected with assigning sense to expressions having the character of proper names is the uncertainty surrounding the question of how to solve the problem of their sense. Some authors are led to postulate a special entity, the individual concept, as separate from individuals themselves. One typical result in this respect is Carnap's effort to extend what he calls intension and extension to expressions of all kinds (Carnap, 1956), i.e. an effort to assign parallels of sense and denotation to all expressions capable of being analyzed semantically. (Carnap calls such expressions designators.)

[9] If we accept the viewpoint that descriptions alone, and not names, can be translated, we must also accept the view that names cannot be replaced by another expression if the sense is to remain the same.

[10] Analytical statements are here meant in the broadest sense, i.e. not merely such statements as 'Living matter is composed of proteins or of other substances than proteins' but also those statements whose truth value depends on advance knowledge of the interpretation or definition of the terms used, for example the statement 'If A is the father of B and B the father of C, then A is the grandfather of C', or 'A volume of 1 dm^3 of water at 4°C weighs 1 kg'.

[11] Popper (1934) speaks in this connection of the problem of demarcation and the criterion of demarcation. He himself offers the suggestion of falsifiability as the criterion of demarcation. Popper's conception, unlike the positivist conception of the Vienna circle, which uses the criterion of verification, is capable of embracing some aspects of the historicity and relativity of any criterion of demarcation.

[12] The most frequent interpretation of general statements makes use of the three-factor theory. This interpretation might be illustrated as follows: 'If any object which is A, is B, then it is C', where A, B and C may be understood as descriptions.

[13] One of the first systematically grounded attempts was that of Carnap (1934) who suggested replacing names by descriptions with time-space co-ordinates. He considered the form of statements with proper names to be more primitive and to correspond to a lower stage of science. Statements containing descriptions with explicitly determined time-space co-ordinates correspond to a higher stage of science. From the Vienna circle's original standpoint the so-called *Protokollsätze* answered these requirements. Their view on the meaning, conception and use of *Protokollsätze*, of course, underwent considerable change in the course of its later development.

[14] Russell (1948).
[15] In linguistically oriented works attention is usually drawn to the fact that for what Quine calls the predicative position it is the verb which is of major significance. This is why wherever the decisive element is a noun or adjective there is always also a verb (e.g. '...*has* a diameter of ---') or at least the copula 'is', 'are', etc. See Quine (1960, p. 96).
[16] Carnap's term 'property' as an object of a conceptual nature is of course distinct, as we shall see later, from the term 'property' as an element of a universe in our communication model.
[17] As far as sentences are concerned this universe is the respective matrix of truth values.
[18] This procedure conflicts with the intentions of extreme nominalism, which, in the analysis of the languages of science, is obliged to abide by languages of the first order. This, however, represents a considerable restriction and does not correspond to the structure of the language of the empirical sciences.
[19] Quine (1953), p. 103.
[20] For a more detailed criticism of Quine's conception of ontic commitments see Tondl (1966b).
[21] This method was pointed out by A. Pap (1962; pp. 126–128).
[22] This distinction corresponds to the difference between one-placed and many-placed predicates.
[23] Hempel (1952).
[24] It can be fairly easily proved that when these requirements are respected in the empirical and experimental sciences we often encounter difficulties. This explains the frequent incertitude as to the inclusion of objects in this or that class in the case of taxonomic classification constructed beforehand.
[25] Attention to these possibilities of reducing many-placed predicates to two-placed has been drawn by Carnap (1950).
[26] This term was introduced by Hempel (1952).
[27] The intervals between the markers in the diagram are not equal. It should be emphasized, however, that the introduction of the concepts of intervals and their measuring in itself presupposes the precisely determined application of quantitative concepts and so too of more accurate measuring. In fact the matter here is simply that the use of a finite number of markers in a given quasi-series does not guarantee that the respective intervals are comparable. This is the case with Mohs' scale of hardness, the Beaufort wind scale and others.
[28] Carnap (1950).
[29] Although the modified comparative predicate A_i' may serve for defining predicates A_i and C_{A_i}, it is in fact a molecular predicate combining components having different properties. It is indeed possible to say that

$$A_i'x, y \equiv \sim (\sim A_i x, y . \sim C_{A_i} x, y),$$

but, as we have shown already, $A_i x, y$ and $C_{A_i} x, y$ as binary relations are not relations of the same kind. Another circumstance which increases doubts as to the transitivity of the relation expressed by the predicate A', is the fact that coincidence must be demonstrable by physical processes (measuring). Menger (1959) has already shown that it is problematical whether physical procedures are capable of guaranteeing transitivity.
[30] Theoretically it is possible to imagine *n*-tuples with quite a considerable number of members. However, with the practical application of quantitative predicates in the empirical and experimental sciences we always find them restricted to a fairly small number. For example, the consumption of a motor vehicle can be determined by a pair of numbers,

namely the number of litres of petrol per n km. Nevertheless this consumption can be determined with greater accuracy by the use of a quadruple of numerical elements, i.e. in m litres of petrol for n km at an average speed of o km per p h. Account may also be taken of the octane number of the petrol used, permitted maximum and minimum speeds and so forth, so that the actual number of numerical elements figuring in the quantitative determination of fuel consumption can be further extended. In metrical and experimental practice, however, it is deemed more expedient to keep this number down or at least to replace some of the elements by standard invariables in all similar quantitative determinations.

[31] Carnap (1950).

[32] A more accurate determination of the degree of hardness of the substance in question can only be achieved using the modified comparative predicate A_1' ('to be harder or as hard'). The substance has the degree of hardness i if it is an element of the class which emerges from the intersection $(\hat{x})\ A_1'x, m_i \cap (\hat{x}) \sim A_1'x, m_{i+1}$.

[33] In the simplest cases this may mean that the intervals can be characterized in the same way. For example, the same amount of heat is required to raise the temperature of a given mass of water from 10 °C to 11 °C as is needed to raise the temperature of the same mass of water from 90 °C to 91 °C. The situation is rather more complicated whenever the metrization of the quasi-series cannot make use of linear dependence: the difference between an IQ of 70 and one of 75 is not comparable to the difference between IQ's of 120 and 125. In this case the differences follow from the assumption that there is a normal (Gaussian) distribution of the properties which we measure by the use of IQ tests.

[34] In the empirical and experimental sciences we sometimes encounter the abbreviation $x=1$ (grams, litres, centimetres, etc.). This does not of course mean identity but merely the equivalence of quantities assignable to both x and m.

[35] We have already drawn attention to this fact in the account we have given of the communication model of cognitive processes designated by the schema

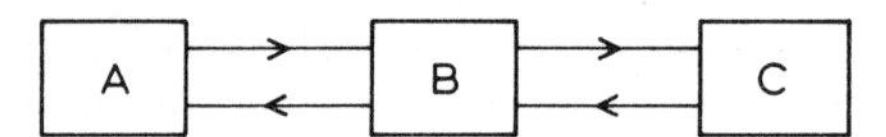

Since we locate the class $\mathscr{M}$ in block B, the universe $\mathscr{X}$ being in block A, we must assume feedback between A and B and so too the familiar cases of retroaction of B on A (such familiar cases are usually demonstrated in the domain of measurement in quantum theory).

[36] Additivity is of course limited by quite a number of factors: the fineness of differentiation of the respective scale factors, the limited adequacy of the scale factors beyond certain ranges of the properties to be quantified, and others.

[37] Campbell (1920, 1928).

[38] Campbell also connected the difference between fundamental and derived measuring with this differentiation. Only extensive properties make fundamental measuring possible, this being then tied to the possibility of an immediate additive operation. He shows, for example, that the measurement of temperature cannot be of the fundamental kind since there exists no process in physics for addition in the case of temperature. He accordingly considers temperature in physics to be an intensive property, just as, for instance, hardness in mineralogy (Campbell, 1928; p. 119).

[39] Cohen and Nagel (1934).

[40] Hempel (1952).

[41] Hempel (1952, p. 76).

[42] In our account of the properties of operation $\circ$ we shall be making use of Tarski's terminology and system of axioms for addition (Tarski, 1961).

[43] Whether the operation $\circ$ is right- or left-invertible is determined by the nature of the predicate A_1, i.e. according as it means 'having the property P_1 in a greater measure than' or 'having the property P_1 in a lesser measure than'.
[44] or $<$, if A_1 denotes 'having the property designated P_1 in a lesser degree', i.e. in the case of left-invertibility.
[45] This isomorphism naturally has limitations which are thrown into relief best if we make use of the communication model, where the class of scale-factors come under block B, while the objects whose properties are to be quantified come under block A. These limitations are very nicely illustrated in the case of measuring in quantum theory.
[46] The emphasis here laid on the fact that at least three kinds of assignment are involved concerns above all aspect (c): on the assumption that the quantification of a predicate requires the assignment of a pair of numerical values, i.e. where $f(P_1) = \langle i, j \rangle$, the assignment meant under (c) is doubled.
[47] The communication model presents certain other possibilities in the analysis of such problems as *thresholds* of discriminability, fineness of discrimination, the capacity of the measuring apparatus, errors in measuring and so forth. However, the analysis of these methodological problems of measuring is beyond the competence of this work.
[48] Russell (1940).
[49] Putnam (1960).
[50] Quine (1953).
[51] This program has been expounded most systematically by Carnap in his *Logischer Aufbau der Welt* (Carnap, 1928). With time, Carnap gradually moderated this program to end with the admission that it is impossible to carry it out with absolute consistency.
[52] Carnap (1936).
[53] E.g. Juhos (1959).
[54] Goodman (1955).
[55] These examples of theoretical predicates are usually interpreted as the products of idealization. Idealization, however, is only one of the ways which lead to the construction of theoretical predicates.
[56] Bridgman (1951).
[57] A critical appraisal of some of these attempts has been given by Hempel (1963). Similar thoughts on the irreducibility of theoretical constructs to empirical predicates are persuasively expounded by Lednikov (1967).
[58] The theoretical predicate 'transitive' is of course a metalanguage predicate with respect to the language of which the signs used are a component.
[59] The term 'correspondence rules' is used by Carnap in particular. A detailed account of them appears in his *Philosophical Foundations of Physics* (Carnap, 1966).
[60] The term 'coordinative definitions' (Zuordnungsdefinitionen) was used by Reichenbach, who in this connection analyzed particularly the relations between geometrical concepts and their physical interpretations.
[61] In a similar way Carnap formulates the criteria of the empirical meaningfulness of any theoretical term in a given theory (Carnap, 1956b). This meaningfulness should be understood as respective not only to the whole theory T, which may be imagined as a set of postulates which define implicitly the theoretical predicates, but also to the aggregate of interpretational ('correspondence' in Carnap) rules C. If S_M is a sentence containing as its only theoretical predicate M, S_k a sentence whose terms are all empirical predicates, and the conjunction $S_M.S_k.T.C$ logically indisputable, then the theoretical predicate M (a component of theory T) is empirically meaningful provided a sentence S_j, which contains only empirical concepts, is implied by the conjunction $S_M.S_k.T.C$ but not by $S_k.T.C$.

[62] What is meant of course is the ideal gas. Since it is assumed that real gases always represent a certain approximation to the ideal gas, operations involving this dependence do also have empirical significance for real gases, on the understanding, of course, that a certain degree of inaccuracy is tolerated. At the same time the fact should be duly respected that the degree of accuracy in physical measuring also has its limitations.
[63] Attention has been drawn to this fact by Carnap (1966).
[64] It is interesting to note that this conclusion was emphasized by Carnap himself (Carnap, 1966; p. 234) who in the thirties had been an advocate of the empiricist programme for building up the language of science.
[65] G. W. Leibniz, *Opera philosophica*, ed. by Erdmann, 1940, p. 94.
[66] We have also demonstrated previously that there are certain difficulties connected with the assignment of denotation in the case of some of the kinds of expressions.
[67] I have indicated the connection between these problems and traditional semantic antinomies in Tondl (1966a).
[68] Particularly good examples of non-extensional contexts are the cases described by Russell as 'propositional attitudes' and nowadays most commonly known as belief-sentences, e.g. 'I think that ...', 'I am convinced that ...', 'I believe that ...' and so on.
[69] Quine distinguishes two principle modes, namely 'use' and 'mention'. In linguistics it is customary to distinguish between when an expression denotes an extralinguistic entity and when it denotes itself, the so-called heteronymous (heterological) and autonymous (autological) uses of the expression.
[70] If we were not to do so we should easily arrive at the utter denial of the possibility of identification. Since it would be impossible to state absolutely everything that might be possible about any object existing independently of the observer or experimenter, we might conceivably apply Wittgenstein's famous slogan to the whole problem. We have in mind the slogan with which he ends his *Tractatus Logico-Philosophicus*, namely: *Wovon man nicht sprechen kann, darüber muss man schweigen.* This may be also taken as the justification for the scepticism which Wittgenstein expresses in respect of Russell's definition of the sign '=' (5.5302) and in respect of any statement on the identity of two things: *Von zwei Dingen zu sagen, sie seien identisch, ist ein Unsinn, und von Einem zu sagen, es sei identisch mit sich selbst, sagt gar nichts* (5.5303). Some of the advocates of logical nominalism also refuse to consider the identity of objects for similar reasons.
[71] The doubts spring chiefly from the viewpoint of extreme nominalism. Since it is expressions of a higher type than zero that are quantified, the problem arises of the ontic commitments of these expressions, i.e. the problem of whether there is not perhaps a hint of the recognition of the existence of the entities to which the expressions relate. These are the reasons why, as we have already shown, many adherents of extreme nominalism in logic dismiss any such approach outright. Those who do not do so while nevertheless considering the identity of objects, introduce procedures which do not require quantification of expressions of a higher type. This is the approach selected by, for example, Goodmann (1951) who defines the sign '=' on the basis of the predicate 'overlap', which in symbolic notation is denoted by $\circ$ and is not transitive:

$$x = y = df\,(\forall x)\,(z \circ x \equiv z \circ y)$$

[72] Particular use is made of various standards and their derivatives as scale factors. The assumption is that quantitative characteristics can be assigned to the scale factors.
[73] Watanabe (1965).
[74] With the procedure to which Watanabe (1965) adheres this question cannot emerge since the distinction of the relevance of the various predicates is not made. This means that it is only the number of coincidences in a selected class of predicates that is decisive.

[75] In the sense in which they are interpreted by, for instance, Carnap (1958).
[76] All the previously mentioned postulates which express, for example, the properties of comparative predicates, should also be treated as meaning postulates.
[77] When it comes to the realization of these procedures in practice we may sometimes be content with results the level of which differs only very slightly from the level attained on the basis of all criteria.
[78] The concept of the 'quality of the solution of a given task' can be variously understood. In the case of bulk statistical tasks it is advantageous to measure the quality of the solution by the average risk attached to the various types of solution.
[79] The whole problem of reduction in scientific procedures, in particular its admissibility, is dealt with in detail in Perez and Tondl (1965).

CHAPTER V

SCIENTIFIC EXPLANATION

1. PROBLEM-SOLVING SITUATIONS AND QUESTIONS IN SCIENCE

(a) *The Concept of 'Problem-Solving Situation' and the Role of Questions in Problem-Solving Situations*

In science, a situation which calls for the explanation of a certain event is a special instance of what we may generally describe as problem-solving situations. The type of situation we have in mind is one where we are aware of something awaiting discovery, something uncertain, undecided or unknown, where we require information capable of improving the level of our knowledge, removing or at least reducing uncertainty, or enabling us to find a suitable solution. In such situations, it is perfectly normal to speak of problems. From this point of view, as we have already shown, scientific cognition may also be described as the solution of problems. In the section on science as cognitive activity we also showed that two aspects must be distinguished in any analysis of problem-solving situations, namely what is known, discovered and decided already, and what remains to be discovered and decided. For these reasons the expression of problem-solving situations calls for the use of two mutually connected components: questions and answers. Intuitively the connection between these two components is obvious from those situations which are typical for activity in science or in the mastering of the results of scientific cognition. When, in the course of his activity, a scientist arrives at at result, e.g. a result of measuring, observation or experiment, which he expresses in the form of statements, the question presents itself to other scientists, and ultimately to the original scientist himself, whether the result achieved is true. This problem-solving situation then leads to the requirement of verifying the statements obtained. An analogous situation is when the results attained have the nature of general statements, such as hypotheses or scientific laws, based on the assumption that they are valid for a limited or assumed universe. In this case it is generally difficult to think in terms of verification, especially if there is no hope of checking on all individual or possible instances of the given universe. Rather do we consider the degree to which a hypothesis is confirmed with regard to the evidence already obtained. A very frequent type of problem-solving situations is that in which we are required to find out which objects (of a class of objects which is defined or at least assumed beforehand) have the desired properties. A slightly different situation is that which arises when we discover a new object and are faced with the task of describing it, discovering its properties (including its relations to other objects), placing it in the appropriate class of objects, and so forth. These cases are sometimes described as tasks of classification or classificational problems. Here, too, the expression of the problem-solving situation calls for the formulation not only of certain statements but also of certain questions.

A special group of problem-solving situations is represented by those which are generally expressed in natural language by a question beginning with 'why'. In the main, these are situations connected with the requirement of explanation, scientific explanation, causal analysis, and so on. Although scientific explanation does include the discovery or derivation of scientific laws or hypotheses to serve as the starting-points for sufficiently reliable predictions, the demands placed on scientific explanation can also be formulated using questions beginning with 'why': 'Why is there an increase in the temperature of the organism in the case of illness?', 'Why do metal objects increase in volume with an increase in temperature?', 'Why were there periodical economic crises in the period of classical capitalism?'. These examples and others like them, of course, do no more than represent a simplication of this kind of problems which are the subject of interest in science. It need at the same time be stressed that the actual verbal formulation is not the main thing. Indeed it is not difficult to show that many formulations of the requirement of scientific explanation which do not use the word 'why' can be reshaped in such a way as to use it. This applies in particular to the requirement of causal explanation ('What is the reason for ...', 'What causes ...') but also to some other kinds of explanation (e.g. 'What is the role of these elements in the given system?', 'What part is played by a given substance in the course of chemical changes?'). For these reasons it will be of some use to investigate the logical and semantic nature of 'why' questions before proceeding to the analysis proper, and also to find out where this kind of question stands in the system of possible question-types generally.

There exists a quite considerable literature on the logical structure of questions.[1] The majority of these works give most attention to the analysis of the logico-syntactic structure of the system of questions. The present work leaves such problems, e.g. those concerning derivation in the system of questions, to one side, concentrating on the role of questions in problem-solving situations. This also means that we shall be respecting the semantically relevant connections of certain questions and answers.[2]

The proposed analysis of questions in problem-solving situations has certain serious limitations which must be respected. The latter are in fact due to the finitistic approach to questions and their role in problem-solving situations. This means that our analysis will be based only on *closed questions.* In practice this in turn means that we adhere much more to the problem of how questions can be answered by a computer or other cybernetic equipment which never has more than finite means at its disposal, than to the problem of how questions are in fact answered by man.[3]

The distinction between open and closed questions is based on the relation of question and answer, or the aggregate of all possible answers, as the case may be.[4] Open questions are generally taken to be those where we are not in a position to give a complete and exhaustive list of possible answers. Giedymin (1964) says in addition to this that in the case of open questions we cannot give either a schema of answers or an effective method of building up permitted answers. On the other hand, closed questions may be taken as those where we are able to supply an exhaustive list of answers, or a schema of answers or effective method of building up the permitted answers. This distinction must always be treated as contingent and relative, since a change in the system which serves as basis may lead to the change whereby a closed question becomes open and vice versa. Such a distinction seems also to be limited in its significance from the historical view of human cognition and its development, which leads to the impossibility of setting up any clear-cut boundaries. It might seem that from the point of view of the infinite and inexhaustive road of the development of human knowledge, all specifically scientific questions are open, since in the future it will become possible to add other unforeseeable answers to the total of known or possible answers to the questions of science as we have them today. However, the real development of human knowledge has involved both exhaustive and finite answers – such as to the question of the possibility of perpetuum mobile – and partial, historically conditioned and incomplete answers which came in time to be completed, corrected or modified.

Hence, in the analysis of problem-solving situations, we must confine ourselves to closed questions and so respect the demands of methodological finitism. This, however, does not mean that the classes of objects (of any kind) with which we operate in a given problem-solving situation cannot be extended to include further objects, in other words it does not mean that constitution or reduction within the framework of these classes is out of the question. If we place the question as to which object, always assuming the object to be one of a certain class of objects, has the desired properties, we do not exclude the possibility that other objects may be included in that class.

Preeminently, open questions are those which are answered by a series of statements which can be comfortably continued at will without the nature of the answer suffering any change. Giedymin (1964) calls these narrative questions. We come across this kind of answer wherever what is called for is, for example, a description understood empirically, an account of the results of observations, or subjective experiences, and so one. The following are typical examples of narrative questions:

'What do you observe in this segment of nature?',
'What impressions do you have from your trip abroad?',
'What do you imagine when you hear this music?'.

In the analysis of closed questions the distinction is generally drawn between what are called *whether-questions* and *which-questions*.[5]

It should be realized that the actual words 'whether' and 'which' are not the main thing in distinguishing the two kinds of questions. Indeed, there are many cases where they do not in fact appear at all. The following are examples of 'whether' questions:

'Is London on the Thames?',
'Does mercury have the properties of metals?',
'Is this alloy corrosive?',

'Is linear programming the best way to solve this particular task of economics?'.

As examples of 'which' questions we may take the following:

'Which is the largest city in France?',
'What is the time?' (i.e. *which* hour of the whole twenty-four),
'When was America discovered by the Europeans?' (i.e. in *which* year ...),
'Who is the author of *Waverley*?' (i.e. *which* writer ...),
'What (*which*) properties does this substance have?',
'How expensive is this apparatus? (i.e. what (*which*) amount of money does it cost?).

From the examples it will be quite clear that the expressions 'whether' and 'which' are not necessary to distinguishing the two kinds of questions. What is the main thing is their respective semantic structure. Furthermore for the semantic analysis of the two types of questions it is important to know the way in which to conceive the relation between the question on the one hand and those statements which comprise answers to the question, or those sentences or expressions which can be reconstructed as the starting basis of the question, on the other hand. (In the account to follow we shall show that the starting basis of 'whether' questions are indicative sentences, and that of 'which' questions are sentence functions, an answer representing the realization of this function.)

Although the majority of works in erotetic logic are confined to the analysis of these two basic types of questions, it is also expedient to distinguish certain other types. For the purposes of application in the sphere of methodology in science, and also for the analysis of teaching processes, the most important type is that of 'why' questions.[6] These are associated as a rule with the demand for explanation, the establishment of causal connections, the justification of a given assertion, and suchlike. As in the previous cases, so here too it is not the word 'why' itself that is essential but the starting basis of the questions. (In this case we shall demonstrate that the basis of these questions is assertion.)[7]

In the next section of this work we shall confine ourselves to the analysis of the heuristic of problem-solving situations associated with 'whether', 'which' and 'why' questions, treating them moreover as closed questions. This is, naturally, not intended as saying that we exclude the possibility of these kinds of questions occurring also as open.

(b) *Problem-Solving Questions Based on Whether-Questions*

One very elementary form of problem-solving situations are those allowing of only two possible solutions, which may be expressed by the word 'yes' and 'no'. Accordingly, questions which occur in such situations are often described as yes-or-no questions. We shall show in due course that these are in effect the most elementary form of whether-questions. The answer to the question 'Is mercury a metal?' is 'Yes, mercury is a metal'. And one answer to the question 'Is this substance a good conductor of electricity?' might well be 'No, this substance is not a good conductor of electricity'. In these and similar questions the starting basis is a sentence in respect of which the pronouncement of the whether-question has the aim of attaining a decision as to the semantic characterization of that sentence, namely a decision as to its truth or falsehood.

The examples given of the yes-or-no type of questions, whose starting basis are sentences awaiting a decision as to their truth value, suggest a certain possible conception of questions based on their reduction to a disjunctive pair of sentences. This conception of questions as a class of sentences joined disjunctively has been worked out by Harrah (1961, 1963). His conception might be expressed as follows: Let S_1 be the sentence 'Mercury is a metal', and S_2 the sentence 'Mercury is not a metal'. Then the question 'Is mercury a metal?' can be expressed formally as

$$(S_1 \,.\, \sim S_2) \vee (\sim S_1 \,.\, S_2).$$

(In this example it is self-evident that $\sim S_1$ is equivalent to S_2.) These reasons also lead to Harrah's name for these questions as *disjunctive questions*.

The notation given here considers only two possibilities, i.e. the 'yes-or-no' type of question. However, there also are questions where it is useful to consider more than two eventualities. The question 'Did white win this game of chess?' allows us to assume 'White won' (S_1), 'White lost' (S_2) and 'The game ended in a stalemate' (S_3). The question can then be expressed, along the lines of Harrah's conception, thus:

$$(S_1 . \sim S_2 . \sim S_3) \vee (\sim S_1 . S_2 . \sim S_3) \vee (\sim S_1 . \sim S_2 . S_3).$$

This way of expressing questions by the disjunctive concatenation of sentences, which is in effect their reduction to a certain chain of sentences which may be considered as the starting basis of the questions or as possible answers, has the disadvantage that it fails to take account of the fact that the relations between these sentences are quite fixed.

We can also readily imagine situations where the answer to a whether-question takes the form 'I do not know whether ...'. Naturally, such an answer does not in fact answer the question, and is, moreover, introduced by an expression which has the character of epistemic modality. Hence it is scarcely possible to place this answer on a level with the ordinary positive or negative answers to a whether-question.[8]

If we ignore the semantically relevant relations between sentences S_1, S_2 and S_3, it would be possible to take as a way of expressing the question an expression which corresponds to all the logically possible state descriptions, i.e. the disjunction of all the given sentences or their negations when

$$(S_1 . S_2 . S_3) \vee (S_1 . S_2 . \sim S_3) \vee (S_1 . \sim S_2 . S_3) \vee \ldots \vee$$
$$(\sim S_1 . \sim S_2 . \sim S_3).$$

It will be apparent that this conception of whether-questions as the disjunctive linking of sentences $S_1, S_2, \ldots, S_n$ where $n>1$, confronts a number of difficulties. Some of these arise out of the situation where the semantically relevant relations between the individual sentences $S_1, S_2, \ldots, S_n$ are not known. These difficulties are to a degree eliminated by Belnap's conception (Belnap, 1963) which, unlike Harrah's, does not reduce questions to sentences. Belnap introduces special question operators by which to differentiate the various kinds of relations between the sentences $S_1, S_2, \ldots, S_n$ which go to make up the basis of the question. In the context of whether-questions Belnap distinguishes these three operators:

$$N(S_1, S_2, \ldots, S_n),$$
$$U(S_1, S_2, \ldots, S_n),$$
$$C(S_1, S_2, \ldots, S_n).$$

He calls questions with the N operator 'non-exclusive whether questions', and the expression $N(S_1, S_2, \ldots, S_n)$, which is the formal expression of the question, might be interpreted as the imperative sentence: 'Say some true sentence from the series of sentences $S_1, S_2, \ldots, S_n$'. It follows from the very name of the operator that the possibility of there being more than one true answer is not excluded, i.e. that if $S_i (i \leqslant n)$ is true, it does not follow that $\sim S_1, \sim S_2, \ldots, \sim S_{i-1}, \sim S_{i+1}, \ldots, \sim S_n$.

Belnap further calls questions with the U operator 'unique alternative whether questions'. The expression $U(S_1, S_2, \ldots, S_n)$, which is the question's formal notation, might be interpreted as: 'Say the only true sentence in the series $S_1, S_2, \ldots, S_n$, all the remaining sentences being false'. In other words the answer to a question with the U operator is the conjunction

$$\sim S_1 . \sim S_2 . \ldots . \sim S_{i-1} . S_i . \sim S_{i+1} . \ldots . \sim S_n,$$
where $1 \leqslant i \leqslant n$.

Belnap calls questions with the C operator 'complete list whether question'. The expression $C(S_1, S_2, \ldots, S_n)$ may be accordingly interpreted as meaning: 'Say all the true sentences in the series $S_1, S_2, \ldots, S_n$, so that all remaining sentences are false'.

The introduction of various operators for whether-questions shows that in the semantic analysis of questions it is possible to take into account certain semantically relevant relations between the sentences which form their basis. However, in Belnap's conception the number of possible relations is reduced to three, which would seem to be a considerable impoverishment and simplification of the actual nature of these relations. It ought also to be borne in mind that semantically relevant relations of this kind represent a kind of apriorism which is already subsumed when the question is pronounced and is then taken for granted in the answering of the question. This is particularly true of the questions which figure in problem-solving situations.

These then are the reasons which lead us to present our own conception of the question in a problem-solving situation, a conception which considers the semantically relevant relations between the sentences that form the basis of the question, and which, like Belnap's conception, introduces a question operator. Unlike Belnap's conception, however, a single operator suffices for all whether-questions.

If (?) is the whether-question operator, then (?) (S_i) means that we are to decide whether sentence S_i is true or false and, in addition, whether or not we are able for one reason or another to give any other decision.[9] If we take as an example the question 'Have you established whether this alloy resists corrosion?', there are the following possible answers: 'Yes, we have established that this alloy resists corrosion' (S_1), 'We have established that this alloy does not resist corrosion' (S_2), 'We have not so far established whether this alloy resists corrosion' (S_3). (It is, of course, quite possible to imagine still more possible answers and so too a greater number of alternatives.) Remaining with these three eventualities: the answer to the original question expresses a decision as to the semantic characteristic of one of the sentences given as those forming the basis of the question. We shall therefore be regarding an answer as an assertion, so that the possible answer in the given problem-solving situation will be $\vdash S_1$ or $\vdash S_2$ or $\vdash S_3$. One characteristic feature of the question in the given problem-solving situation is that from the moment when the question is uttered due account should be taken of a certain apriorism, i.e. certain semantically relevant relations between S_1, S_2, and S_3 which are fixed beforehand. We shall further be assuming that $\mathfrak{P}$ is the class of all the meaning postulates which express all the semantically relevant relations existing between the sentences which comprise the basis of the question. This means that when we come to express the given question we assume in advance, on the one hand, sentences which make up the basis of the question, and on the other hand, the respective meaning postulates; we indicate this by

$$[S_1, S_2, S_3, \mathfrak{P}].$$

It is thus obvious that we take into account a broader spectrum of the possible relations between the sentences which make up the basis of a question than is the case in Belnap's conception. The question 'Have you established whether this alloy resists corrosion?' may then be recorded schematically as

$$(?)\,(S_1)\,[S_1, S_2, S_3, \mathfrak{P}].$$

The expression in square brackets fixes the information which is assumed when the question is uttered. This also means that the operator does not relate to the expressions in the square brackets. Hence the distinction must always be made between whether a certain sentence occurs in the actual formulation of a question (i.e. in front of the square brackets), as part of the information assumed when the sentence is uttered (i.e. in the square brackets), or in an answer, which we always mark with the sign ⊢. These are in effect three different modes of using sentences.[10]

The schematic notation given above, taking into account a certain apriorism as it does, is in fact the expression of a problem-solving situation connected with a whether-question. This kind of notation also gives a much more adequate expression of the actual contexts in which questions occur. For if we pronounce a question, we generally take for granted certain information as known (for both questioner and questionee). This may take the form of a tacit assumption or may be explicitly expressed in the formulation of the question.

From the point of view of the conception of yes-or-no questions, a certain objection may be raised against the formal notation which also respects the apriorism. Let us revert to our example of sentences S_1, S_2 and S_3 and their interpretations. The objection might be formulated thus: if we get the answer 'We have not so far established whether this alloy is resistant to corrosion', the problem-solving situation in effect remains unsolved. The original uncertainty, and hence too the original question, remains in force. In line with this objection, therefore, the only legitimate form of a problem-solving situation associated with a whether-question is

$$(?)\,(S_1)\,[S_1, S_2, \mathfrak{P}],$$

where $\mathfrak{P}$ has the form $S_1 \equiv \sim S_2$ and $S_2 \equiv \sim S_1$. This objection is justified up to a point in the analysis of questions and answers where we take the questions in isolation. However, it is not justified in the analysis of a question in a problem-solving situation, since in the overwhelming majority of situations there is little to be gained from reducing the components of the situation which we take into account when uttering the question and which are expressed in the notation in the square brackets, merely to the alternative of a sentence and its negation.[11] Let us take a look, in view of this, at the simple question: 'Did white win this game of chess?' In line with the objection there would only be two possible answers, namely 'White did win this game of chess', and 'White did not win this game of chess'. In fact the class of possible and practical answers in the solution of any problem-solving situation is greater and may include such information as, for instance, that the game ended in a stalemate or that one side resigned. In theory the reduction of the class $\{S_1, S_2, \ldots, S_n\}$ to the pair $\{S_1, S_2\}$ where S_2 is the complement of S_1 is naturally not precluded, but it need not be always expedient as is indeed testified by everyday practice in the use of language.

The solution of the heuristic of problem-solving situations connected with a whether-question usually requires in practice that we respect the fact that when the question is uttered certain information is taken for granted beforehand; and this information has the character of the semantically relevant relations between the sentence or sentences which form the basis of the question, and other sentences. The same information also gives the problem-solving situation a closer specification, although in the interests of simplicity it need not always be stated explicitly. We may demonstrate these circumstances on the basis of the question: 'Did you establish in your experiments yesterday whether this substance contains chlorine?'. It is clear that if we wish to take account of the entire problem-solving situation it is insufficient to confine ourselves to this question and its sentential basis alone. The very act of putting the question assumes that we take for granted the information contained in these sentences: 'You are a chemist and you make experiments', 'Yesterday you were doing some chemical analyses', 'You were analyzing this substance yesterday' and so on. Thus the question need not be answered merely by the confirmation or denial of the sentence which forms its basis. We are normally quite willing to consider as a satisfactory answer the following for instance: 'It is true that I did some chemical analyses yesterday, but not of this substance'.

These and similar cases show that in practice we must allow for quite a large number of alternatives which stem from the apriorism relevant to the given question. If there are n alternatives among which there exist certain fixed relations, then a problem-solving situation associated with a whether-question can be expressed in formal notation as

$$(?)\,(S_i)\,[S_1, S_2, \ldots, S_n, \mathfrak{P}], \quad \text{where} \quad 1 \leqslant i \leqslant n.$$

From the point of view of the logico-syntactic structure of the language in which the problem-solving situation is formulated, it need not be assumed of necessity that $S_1, S_2, \ldots, S_n$ are merely atomic sentences. This also means that S_i, which is bound by the question operator, might well also be a molecular sentence.

The expressions which, in our notation for a problem-solving situation, have appeared in square brackets may also be of great significance in *assessing the meaningfulness of the question* and so too the justification of a given question as used practically in problem-solving situations.[12]

The expression in square brackets which fixes the information which is assumed when the question is uttered, may also include what in logical and logico-semantic literature is usually described as 'ontic commitments',[13] whereby the latter may in turn contain criteria of the question's meaningfulness. If we ask 'Is Paris the largest city in France?' we assume (1) that there exists a city called 'Paris', and (2) that this city is in France. If the individual constant a_1 denotes Paris, predicate P_1 the property 'is the largest city in France', and predicate P_2 the property 'is in France', then $\mathfrak{P}$ also includes the assumption that

$$(\exists x)\,(x \text{ is identical with } a_1)\,.\,P_2 a_1\,.$$

We naturally assume certain other sementically relevant relations such as those between the properties designated P_1 and P_2.

An example of a question which we cannot accept as meaningful is a modification of Russell's famous example: 'Is the present king of France bald?'[14] Writing this question in the notation $(?)\,(P_1\, a_1)$, then in terms of Harrah's conception we should have the disjunctive question

$$(P_1 a_1\,.\sim P_2 a_1) \vee (\sim P_1 a_1\,.\,P_2 a_1),$$

where the predicate P_1 is 'to be bald' and P_2 'not to be bald'. (It again holds of course that $(\forall x)\,(P_1 x \equiv \sim P_2 x)$.) Using the sign $\vdash$, then the possible answers to the question ought to be either $\vdash P_1 a_1$ or $\vdash P_2 a_1$. It is quite evident that we cannot think in terms of the truth of either for the simple reason that at present there exists no person whom we can describe as 'the present king of France'. In order for a question $(?)\,(P_1 a_1)$ to make sense it must be demonstrated, in this case, that the person designated as 'the present king of France' whose baldness or otherwise we are asked to ascertain, really exists. Since this cannot be demonstrated the sentence is lacking in sense.

Naturally there are many cases where the confirmation or refutation of a question's meaningfulness is not so simple, as is clear from the examples given earlier. At all events the explicit formulation of the function $\mathfrak{P}$, which also includes the ontic commitments, may be useful when we come to the semantic substantiation of the meaningfulness of a question.

Whether-questions may also be formulated in such a way that from the very formulation it is clear that we consider a certain sentence or sentences to be true. Let us consider the following questions from this point of view. They are of course questions uttered in problem-solving situations. If a doctor asks 'Has the patient's temperature gone down?', it is clear from the formulation that the sentence 'The patient had a temperature' is considered as true. If we ask, in a particular factory, whether they are still having difficulties with the supply of raw materials, we assume that such difficulties have in fact existed in that factory. Questions of this kind are generally described as *suggestive*.[15] From the point of view of the conception of problem-solving situations from which we started out, the information assumed or suggested by a suggestive question might be considered as a component of the meaning postulates $\mathfrak{P}$.

At this point we might outline some of the applications of problem-solving situations connected with whether-questions. The latter occur in problem-solving situations where the main thing is decision-making as to the truth of certain sentences, the confirmation of hypotheses by the evidence available so far, the confirmation of scientific laws by experiments which have been carried out, and so forth. In the examples which are to follow we start from the basic form of whether-question problem-solving situations, expressed in the notational form

$$(?)\,(S_i)\,[S_1, S_2, \ldots, S_n, \mathfrak{P}], \quad \text{where} \quad 1 \leqslant i \leqslant n\,.$$

Let S_i (or one of the sentences $S_1, S_2, \ldots, S_n$, as the case may be) be a singular synthetic sentence. It may represent an unequivocal and definitive decision as to the truth of S_i (or one of the sentences $S_1, S_2, \ldots, S_n$). In this case we have a *vertificational problem*. The natural sciences quite commonly find themselves faced with the task of verifying a sentence as to whether a certain object has this or that property, is a conductor of electricity, contains such and such substances, or has certain given chemical or physical properties. If it is to be established, for example, whether an object a_1 has the property designated P_1, certain operations must be carried out in order for it to be possible to decide whether $\vdash P_1 a_1$ or $\vdash \sim P_1 a_1$. Speaking in general terms, in a question $(?)\,(S_i)\,[S_1, S_2, \ldots, S_n, \mathfrak{P}]$ the decision as to the truth of S_i (or of one of the sentences

$S_1, S_2, \ldots, S_n$) must be based on a sentence S_o which, however, is not an element of the class $S_1, S_2, \ldots, S_n$. In the case of empirical verification the sentence or class of sentences S_o is the expression of the results of observation, measurement and experimental procedures etc., which decide as to the answer to the given question and are capable of verifying (or otherwise) the sentence which forms the basis of the question.

The sentence S_i in the question (?) (S_i) [$S_1, S_2, \ldots, S_n, \mathfrak{P}$] may have the character of a universal sentence and express a scientific hypothesis or law. In this case we speak of the *confirmation* of a hypothesis, not of verification, particularly if we are not in a position to examine in turn all the possible instances of the universe to which S_i relates. The confirmation of a hypothesis by means of the evidence available, i.e. by means of the metrical and experimental results obtained, may be made in varying degrees which might, if need be, also be expressed quantitatively, i.e. by some method of mathematical statistics, or in terms of probability logic.[16]

If S_i in the given question takes the form of the general sentence $(\forall x)\,\mathrm{P}x$, where $x \in \mathscr{X}$ and if it is possible to go through all the elements of the class $\mathscr{X}$ in turn, then the possible answers may take the forms:

$$\vdash (\forall x)\,\mathrm{P}x,$$
$$\vdash (\forall x) \sim \mathrm{P}x,$$
$$\vdash (\exists x)\,\mathrm{P}x,$$
$$\vdash (\exists x) \sim \mathrm{P}x.$$

The last two of these would also be possible even were it possible to investigate only some of the elements of the class $\mathscr{X}$.

A slightly more complicated situation arises if the sentence S_i has the form of the general sentence $(\forall x)\,(\mathrm{P}x \rightarrow \mathrm{Q}x)$. In a case like this we come up against the familiar properties of material implication known also as the paradoxes of material implication if the class $(\hat{x})\,\mathrm{P}\,x$ is empty. A question where S_i does have the form in question can only be meaningfully placed under the condition that $(\hat{x})\,\mathrm{P}\,x$ is a non-empty class and that $(\hat{x})\,\mathrm{P}\,x \subset (\hat{x})\,\mathrm{Q}\,x$.

A very important set of applications of problem-solving situations connected with whether-questions are those where we are to decide whether to accept or reject one of a pair of contradictory hypotheses. In practice this means that we look for evidence capable of confirming or refuting one of the contradictory hypotheses. There are many familiar cases of this both in the history of science and in modern scientific practice. One was that of the geocentric and heliocentric conceptions of the universe, others the contradictory hypotheses on the nature of electricity, light etc.

Let there be a problem-solving situation where it is to be decided whether S_j or S_k, where S_j and S_k are two different hypotheses. Now let the sentences S_o and S_p be possible items of evidence, i.e. results of observation, measurement or experiment, which can verify or at least confirm to a high degree the two hypotheses such that it holds that

$$S_j \rightarrow S_o,$$

and

$$S_k \rightarrow S_p.$$

In the solution of problem-solving situations of this kind the relation between the two hypotheses is material. As far as this relation is concerned, two types of situations can be discerned:

(a) In the first case the matter to be decided is whether S_j or S_k, though at the same time we know that the two hypotheses are incompatible. This means that the confirmation of S_j simultaneously means the refutation of S_k and vice versa. Cases of this nature have traditionally been described as *experimentum crucis*. This was, for example, the case of the relation between the geocentric and heliocentric hypotheses. The solution of the question whether S_j or S_k, might then proceed as follows:

First it must be shown that S_j/S_k. If we know that

$$S_j \rightarrow S_o,$$

and

$$S_k \rightarrow S_p,$$

and if we discover that $\sim S_p$, then we can conclude that $\sim S_k$.

This conclusion – on the understanding that the first assertion in the procedure has been proved – justifies the acceptance of the hypothesis S_j. It is evident that this procedure is based on the *modus tollendo tollens*, though of course this is not the only conceivable procedure. It must be emphasized that in procedures of this kind it must be shown not only that the inferential sequence selected (which need not be the one given in our example) is admissible, but also, and chiefly, that the two hypotheses are incompatible.

(b) The second type of situation occurs when it is to be decided whether S_j or S_k, but where at the same time it cannot be demonstrated that S_j/S_k. In other words the two hypotheses need not be incompatible, so that confirmation in a high degree of one or other of the hypotheses does not mean the complete refutation of the other. For the same reason the fact that available evidence justifies us in accepting one of the hypotheses does not justify us in rejecting the other. The history of science and the problems of modern scientific methodology again provide numerous examples of this kind or problem-solving situation. One case is that of the emissive or corpuscular theory of light relative to the wave theory. Both conceptions could be backed up by empirical or experimental confirmation and yet it was not possible to prove simultaneously that they were incompatible hypotheses. Thus in cases similar to the problem-solving situation mentioned here it is not possible to apply the ordinary procedure connected with the *experimentum crucis*, where the confirmation of one of the alternative hypotheses simultaneously means the refutation of the other and vice versa.

(c) *Problem-Solving Situations Connected with Which-Questions*

In drawing the distinction between problem-solving situations connected with whether-questions and those connected with which-questions, some authors consider the number of alternatives, from among which the decision is to be made, to be essential. Harrah (1963, p. 32), to quote one of them, points out that in whether-questions the decision process is based on a small, or at least finite, number of alternatives. (In the simplest case – (?) (S_i) – there are two alternatives, namely $\vdash S_i$ or $\vdash \sim S_i$.) In contrast, which-questions involve quite a high number, or even an infinite number of alternatives. These are the reasons which lead Harrah to call the former 'multiple-choice questions' and the latter 'fill in the blank questions'.

The kind of differentiation which accentuates the number of alternatives from among which a decision is to be made, although possibly of great significance for the psychological assessment of a given problem-solving situation, does not, I think, capture the essential difference between the two kinds. Although problem-solving situations connected with which-questions do generally involve a relatively considerable number of alternatives, it is not this that is the main thing. In my opinion the fundamental differences are tied up with the role of variables in either kind of problem-solving situation.

Variables are not absolutely necessary in the notational expression of whether-questions, but they are needed in the formalization of which-questions. If the predicate P_1 denotes the property 'is the highest mountain in France', then the question 'Which is the highest mountain in France?' means that we are looking for the individual which *fulfils* the sentential function P_1x. If we ask which of our friends has visited Istanbul, or who knows how to solve differential equations, etc., we may also answer that such and such a person fulfils the given sentential function.

In the schematic notation of questions of this kind we might proceed thus: Since in fact we are to decide which individual fulfils a given function, we may take the sentential function as the basis of these questions. We then introduce the operator $(?x)$ which may be interpreted as 'Which individual fulfils the sentential function P_ix'. Hence the elementary form of such questions is

$$(?x)\, P_ix .$$

We shall call questions of this type 'which individual questions'.[17]

The examples given so far have assumed a sentence function with a single argument as the basis of a which-question. In general, of course, it may be a function with more than one argument, i.e. $P_ix, y, z \ldots$, where it is to be decided which individuals fulfil the function. Nor can we exclude as meaningless questions which lead to the answer 'No individual fulfils the function ...'. Since we are dealing here with decision-making as to the fulfilment of a (sentential) function, we can admit as meaningful questions not only those whose answers give a full list of the individuals which fulfil the given function, but also those which state which individuals or which single individual fulfils the given function, or state that no individual fulfils the function.

We shall proceed now from the simple analysis of which-questions to that of problem-solving situations connected with them. In the case of the latter the matter is no longer whether just any individual fulfils the function, but of deciding which individual of a class, which has been defined in a particular way, fulfils the given function. If we ask which pupil in a class has won a distinction, which is the highest mountain in France, what German rivers flow into the Rhine, and so forth, we give at the same time a specification of the range of variability of the individual variable x, since we are in effect asking which element of the class of all pupils in the class, which element of the class of mountains, etc., respectively. This specification need not always be clear from the actual formulation of the question. If we ask, for example, who has satisfied certain prescribed conditions, the delimitation of the class depends on the overall context, e.g. the circle of people we are addressing. Thus with 'which individual' questions we not only operate with the sentential function $P_i x$, or look for the element of the class $(\hat{x}) P_i x$, but also with the class $(\hat{x}) P_j x$ which specifies the range of variability. We further assume certain fixed relations between the classes $(\hat{x}) P_i x$ and $(\hat{x}) P_j x$ which we express by the sign $\mathfrak{P}$. The most usual and common version of $\mathfrak{P}$ seems to be the inclusion of one of the classes, namely,

$$(\hat{x})\ P_i x \subset (\hat{x})\ P_j x.$$

Now we can express a problem-solving situation in notational form as:

$$(?)\ (P_i x)\ [x \in (\hat{x})\ P_j x,\ \mathfrak{P}],$$

where again the expression in square brackets fixes the information which is taken as given when the 'which individual' question is uttered.

We have already indicated that 'which individual' questions are a sub-group of which-questions. The questions which we might describe as 'which predicate' questions may be considered as an independent group of which-questions. However, in natural languages the distinction between these two groups of which-questions is often far from apparent. From the point of view of the analysis of problem-solving situations connected with which-questions two groups of situations can nevertheless be distinguished: (a) If certain properties are known, or if such properties have been exactly determined or defined, one frequent task is to discover what objects have the properties. In technological practice, an example might be the search for materials having the required flexibility, stability, or resistance to exterior influences. (b) Somewhat different are the situations where it is an object that is known and the task is to discover what properties it has or to what class of object it belongs and so forth. The first group represents problem-solving situations connected with 'which individual' questions, and the second those connected with 'which predicate' questions.

The second group may be recorded in the notation as

$$(?\mathrm{P})\ \mathrm{P}a_1 \qquad [\mathrm{P} \in \mathscr{P}],$$

where (?P) is the operator 'which predicate' and $\mathscr{P}$ is the class of predicates to be taken into account, i.e. $\mathscr{P} = \mathrm{P}_1, \mathrm{P}_2, \ldots, \mathrm{P}_n$. It goes without saying that this solution assumes an explicit knowledge of the elements of the class $\mathscr{P}$. When we come to ask what properties a given substance has, there are two kinds of situation which we may distinguish: (a) The set of properties, e.g. the chemical properties which may be involved, is precisely specified, in which case we speak of tasks of identification. (b) The set of properties, while being specified, can nevertheless be extended by further properties, or have some of its properties left out.

The two groups of which-questions have certain common features connected with the conjunctive nature of the answers. In the case of the question as to which individual of a given class has a certain property, i.e. the question $(?x)\ (\mathrm{P}_i x)\ [x \in (\hat{x})\ \mathrm{P}_j x,\ \mathfrak{P}]$, we must assume that the class $(\hat{x})\,\mathrm{P}_j x$ is a finite non-empty class.[18] If, for instance, $(\hat{x})\,\mathrm{P}_j x$ is a class of objects designated $a_1, a_2, \ldots, a_n$, the answer assumes the selection of those members of the class that fulfil the function $\mathrm{P}_i x$. Hence the answer has the character of a conjunction, namely

$$\vdash \mathrm{P}_i a_1 . \mathrm{P}_i a_2 . \ldots . \mathrm{P}_i a_n,$$

which naturally omits those elements of the class $(\hat{x})\,\mathrm{P}_j x$ which do not fulfil the function. Analogously in the case of a problem-solving situation connected with a 'which predicate' question, i.e. in the case of (?P) $\mathrm{P}a_1\ [\mathrm{P} \in \mathscr{P}]$, the answer is the selection of the elements of the class $\mathscr{P}$ which pertain to object a_1, i.e.

$$\vdash \mathrm{P}_1 a_1 . \mathrm{P}_2 a_1 . \ldots . \mathrm{P}_n a_1 .$$

Having presented this general analysis of the logico-semantic structure of which-questions, we shall now proceed, as in the case of the preceding type of questions, to the analysis of the criteria of their meaningfulness. Some authors hold[19] that from the semantic point of view the only questions that are justifiable are those which admit of at least one positive answer. This means, in the light of this conception, that such a which-question is meaningful only in a problem-solving situation $(?x)\ \mathrm{P}_i x$ $[x \in (\hat{x})\mathrm{P}_j x,\ \mathfrak{P}]$ if $(\hat{x})\mathrm{P}_i x$ is a non-empty class.

However, our experience in everyday life and in research activity teaches us quite clearly that there are very frequent cases where questions lead to the answer that no object of an indicated class of objects has the required property. Questions such as 'What substance is harder than diamond?' 'What mountain is higher than Everest?', and so on, are unquestionably meaningful in spite of the negative answer which must follow ('No substance ...', 'No mountain ...', etc. respectively). This shows that the given criterion of meaningfulness is too severe since it precludes questions which we very often encounter in everyday intercourse, pedagogical practice and research work. Moreover, this criterion does not cover all the situations that may arise. It is expedient to distinguish the following situations:

(1) The class $(\hat{x})P_i x$ is *L*-empty.[20]

(2) The class $(\hat{x})P_i x$ is *F*-empty, but not *L*-empty.

(3) The class $(\hat{x})P_j x$ is empty.

Of particular importance for the solution of criteria of the meaningfulness of which-questions is the difference between (1) and (2). Since we know quite definitely – our knowledge being confirmed not only empirically but also theoretically – that diamond is the hardest substance, the negative answer follows naturally from this knowledge. If knowledge of this kind is given beforehand, there is little point in placing the question in any like fashion. Thus of particular significance from the methodological point of view are the discoveries that lead to the demonstrable confirmation that the class $(\hat{x})P_i x$ is *L*-empty. Of course, this fact need not always be clear from the very outset. We know from both the history of science and contemporary activity in science how important are the discoveries, which are analogous to the theoretical verification of an assertion, that no *perpetuum mobile* exists, that no body can have a greater speed than that of light, and so on. These and similar items of knowledge, which are usually based on the knowledge of scientific laws, accordingly make up a notable apriorism which is important in deciding as to the meaningfulness of questions.

Of a slightly different nature are problem-solving situations where the class $(\hat{x})P_i x$ is *F*-empty. For instance, if we ask our colleagues at work – and whom this involves is precisely defined – who knows how to solve differential equations, we may obtain the answer 'none' or the answer 'some' or even 'all'. While in case (1) above, i.e. that of asking what substance is harder than diamond, or what device has the properties of a *perpetuum mobile*, the existence even of a single element of the class $(\hat{x})\,P_i x$ is excluded beyond possibility, in case (2) the existence of some such element is not precluded. It is, for instance, possible to discover which antibiotic (from a given finite class of antibiotics) is effective against a given disorder. Here too the answer 'none', 'some' or 'all' may occur. A situation where a class $(\hat{x})P_i x$, while being *F*-empty, has not been theoretically proved to be *L*-empty leads to some important consequences. If we have found that the class $(\hat{x})P_i x$ is *F*-empty, and if we nevertheless desire a positive answer, this is a stimulus for extending the class $(\hat{x})\,P_j x$. In other words the discovery that $(\hat{x})P_i x$ is *F*-empty leads to the requirement that we leave class $(\hat{x})P_j x$ in favour of class $(\hat{x})\,P'_j x$, where $(\hat{x})\,Px \subset (\hat{x})\,P'_j x$. The need to expand the original class can be easily demonstrated on the above example with antibiotics. Arriving at the answer that no antibiotic is effective we are obliged to search for others, i.e. to extend the original class. Thus the discovery that class $(\hat{x})\,P_i x$ is only *F*-empty, far from precluding other questions, even stimulates them.

The exact differentiation of problem-solving situations in which $(\hat{x})\,P_i x$ is *L*-empty and those in which it is merely *F*-empty is of great importance methodologically for all areas of application. Having discovered that we are dealing with an *L*-empty class, e.g. that there cannot exist any technical apparatus with the properties of a *perpetuum mobile*, nor any substance with a temperature lower than $-273\,°C$, nor any absolutely noiseless information channel, nor any computer with an infinite memory, etc., then we must fundamentally reorientate any further paths of discovery. As a rule any such conclusion is the result of theoretical analysis and not merely of empirical activity. A negative answer thus has definitive importance here and means that it is pointless to ask any further questions in the framework of the same problem-solving situation.

The situation where the class $(\hat{x})\,P_i x$ is only *F*-empty is different. In this case the negative answer need not be definitive and the possibility of looking for other objects remains, i.e. there is still the possibility of extending the original class so that one of its elements will fulfil the function $P_i x$.

The situation where the class $(\hat{x})P_j x$ is empty presents quite a different case. Even the intuitive viewpoint is enough for it to be quite clear that it is pointless to ask which element of an empty class has a certain previously defined property. Accordingly problem-solving situations of the type

$$(?x)(P_i x)\quad [x \in (\hat{x})\, P_j x,\quad (\hat{x})\, P_j x = 0],$$

are quite foreign to the empirical sciences.

The applicability of a problem-solving situation $(?x)(P_i x)\,[x \in (\hat{x})P_j x, \mathfrak{P}]$ relates in particular to tasks connected with the decomposition of classes (in our notation the class $(\hat{x})\, P_j x$) and with the selection of objects having a certain required property (in our notation objects which fulfil the function $P_i x$). The problems connected with the selection, in particular its optimization and economization, are of importance in many scientific, pedagogical and practical tasks. However, the class in which we wish to make the selection is not always easy to cope with. In practice, when we come to make the selection, we are obliged to respect the finite nature of the means at our disposal, the limitations in the distinguishability of the objects, the finite time-limit for making the selection, and so on. Hence it is of particular importance to keep the number of steps in the process of selection down to a minimum and to organize the process in such a way as to select the particular elements of the class according to their relative relevance. In which-question problem-solving situations – diagnostic procedures, factor analysis, the search for materials having certain desired properties, the search for objects which fulfil a certain function, or whatever – it is also usually necessary to respect certain pragmatic criteria, namely goals, demands placed on accuracy, the quality of the solution, etc.

(d) *Why-Questions and the Problem of Scientific Explanation*

The literature which has so far been devoted to the analysis of questions has paid little attention to why-questions. Fragmentary allusions to their importance are more likely to be found in works dealing with the problems of scientific explanation. Generally speaking, however, the latter rarely go into the logico-semantic nature of this type of question in any great detail.

As in the other types of question, here too it is not the actual verbal form or the explicit occurrence of the word 'why' that is decisive. Moreover, quite different verbal formulations may be selected for equivalent or very similar instances of one and the same question, for example:

> 'How can it be explained that metal objects increase in volume with an increase in temperature?',
> 'What is the cause of the increased volume of metal objects at higher temperatures?',
> 'Why do metal objects increase in volume with an increase in temperature?'

For the sake of giving a more precise indication of how this kind of question differs from whether- and which-questions, let us once more recall what is actually to be decided in the answers to these kinds of question. If we judge questions from the informational standpoint, they are the expression of uncertainty, doubt, etc. [We ignore here all formulations which, though they have the external form of questions, are not in fact questions from the semantic point of view on the grounds that, for instance, they are merely modified forms of statements, or that they contain the answer, and so forth.] A true answer, therefore, represents a lowering of the degree of doubt or uncertainty. Viewed in this way, it is then a legitimate question to ask what kind of uncertainty or doubt is lowered or eliminated by a true answer to a why-question.

If we ask 'whether', i.e. $(?)(S_i)$, the uncertainty relates to the truth value of the sentence S_i or of all the sentences that make up the basis of the question, as the case may be. A which-question shifts the same problem into a more general sphere – that of decisions relating to the fulfilment of the sentential function. This difference is due *inter alia* to the fact, as we have already demonstrated, that in which-questions we assume there are variables.

In contrast to this, the characteristic feature of a why-question, irrespective of whether the little word 'why' appears in the question's actual formulation or not, is that the sentence basis of the question is in itself a true sentence. To be more precise, we are not at all in doubt as to the truth of the sentence basis, which in effect we *assert* as true. So the basis of a why-question is an *assertion*. To return to the examples of why-questions that we have just given: the basis of all three formulations is the assertion 'Metal objects increase in volume with an increase in temperature'. Hence why-questions do not as a rule have the aim of deciding as to the truth of the sentences which form their bases. On the contrary their truth is confirmed and is not then an object of doubt. The matter at stake here is how to justify the assertion.

We often encounter in everyday speech another kind of why-question where the difference is in that the basis is not presented as a true assertion, but merely as any other assertion of something possible, of something that might be materialized, or of a certain viewpoint or opinion. If we ask, for example, 'Why should you like Rome?', 'Why is X convinced there will not be a third world war?', 'Why would the bridge collapse in the case of a greater load than that permitted?', we need not be assuming that the event which is asserted has in fact happened. This same group also includes why-questions relating to future time, such as 'Why won't that building be completed next month?', or 'Why will there be a drop in the birth-rate with an increase in the level of civilization in the developing countries?'. Why-questions of this kind have been called by Hempel 'epistemic or reason-seeking why-questions' in contrast to the foregoing type where the basis is taken to be a true assertion. The basic type of why-question is described by Hempel as 'explanation-seeking why-questions'.[21] This distinction between why-questions which demand an explanation and especially an explanation of an actual event by means of scientific laws and causal connections, and why-questions which only ask for a rational justification or set of reasons why a certain assertion should be accepted, should be clear, especially if we consider the above examples, which have been specially chosen to underline the difference. However, in view of the day-to-day use of why-questions, it will scarcely be possible to find a sharp dividing-line.

In the assessment of the logical and semantic nature of the two kinds of questions involving 'why', one fact of particular importance is that the answering of explanation-seeking why-questions is usually the model way of answering only reason-seeking why-questions. In this connection Hempel quite rightly points out that any adequate answer to an explanation-seeking why-question is also a potential answer to a reason-seeking why-question, but not the converse.[22]

In the remainder of our account of why-questions in problem-solving situations we shall be leaving reason-seeking questions (in the sense outlined above) to one side, concentrating merely on why-questions of the first kind. This means that we accept the assumption that the basis of a why-question is a true assertion. Having agreed on this viewpoint as regards the basis of why-questions, we shall introduce a special why-question operator to denote them. We shall use the sign $(? \vdash)$ as the operator for explanation-seeking why-questions in order to indicate that we do not doubt the assertion that is the basis of a given question. If $\vdash S_i$ is this assertion, then we may express the why-question in our notation as

$$(? \vdash)\,(\vdash S_i)\,.$$

Let us now take a closer look at what we are actually asking for when we ask a why-question. We have already said that if we ask, for example, why wood floats on water, we do not doubt that this is in fact the case. We are as a rule asking for the explanation of the fact to which S_i refers. By explanation we understand the procedure which leads to the extra-logical substantiation of the given assertion. The substantiation of the fact that we admit S_i as true and therefore that we are justified in $\vdash S_i$ must be sought in other assertions and in the relations of the latter to $\vdash S_i$. The answering of the question $(? \vdash)\,(\vdash S_i)$ is therefore rooted in the conjunction of the following procedures:

(a) Finding other statements which we can assert in support of $\vdash S_i$;

(b) Finding the relations or procedures which enable us to prove $\vdash S_i$ on the basis of the assertions of the statements sub (a). (We shall show in due course that these relations are usually understood as relations of inference.)

The linking of the two procedures may be illustrated on the following example:

Question: 'Why has the metal bar increased in volume?' $(? \vdash)\,(\vdash S_i)$.

Answer: (a) 'Because the metal bar has been heated to a higher temperature' $\vdash S_j$ and (b) 'Because whenever a metal bar is heated to a higher temperature it increases in volume' $\vdash (S_j \rightarrow S_i)$.

In this example S_i, which is the basis of the why-question, is at the same time the conclusion of an inferential procedure, known as the *modus ponens*. So the question $(? \vdash)\,(\vdash S_i)$ is answered by: $\vdash S_j, \vdash (S_j \rightarrow S_i)$.[23]

This version represents no more than the simplest form of answering the question $(? \vdash)\,(\vdash S_i)$, the assertion S_i being justified only on the basis of $\vdash S_j$. The same procedure may also be expressed in a general form. We introduce the sign Inf [---, ...] to mean that the expressions '...' can be inferred from the expressions '---'. To answer a question $(? \vdash)\,(\vdash S_i)$

means finding a (finite) class $\{S_j, S_k, \ldots, S_n\}$ so that $\mathrm{Inf}[\{S_j, S_k, \ldots, S_n\}, S_i]$.

We shall now proceed to show some of the more important kinds of problem-solving situations in which why-questions occur. The simplest and easiest of solution are those where the inferential procedure Inf [---, ...] is taken as given or known. These are primarily various kinds of deductive inference.

In the following two examples, Inf [---, ...] is assumed to be the *modus ponens* and the substitution rule respectively.

(1) Take the simple question: 'Why is there mud?'. The answer is: 'Because it has been raining, and whenever it rains there will be mud.' In the notation this is

Question: $(?\vdash)(\vdash S_i)$.
Answer: $\vdash S_j, S_k, \ldots, S_n$ and
$\vdash(\{S_j, S_k, \ldots, S_n\} \rightarrow S_i)$.

We usually understand the class of sentences $\{S_j, S_k, \ldots, S_n\}$ as a conjunctive class. In all like cases we treat the circumstances denoted $\{S_j, S_k, \ldots, S_n\}$ as the necessary conditions without which the fact denoted S_i cannot be explained.

(2) In explanation we often operate with general sentences. We might ask, for example: 'Why does wood (a_1) float on water (P_1)?'. The answer operates with the general sentences: 'Because it is true of all substances with a specific gravity less than 1 (P_2) that they float on water' and 'Because wood is an element of the class of substances with a lower specific gravity than 1'. In our notation this is

Question: $(?\vdash)(\vdash \mathrm{P}_1 a_1)$
Answer: $\vdash(\forall x)(\mathrm{P}_2 x \rightarrow \mathrm{P}_1 x)$ and
$\vdash a_1 \in (\hat{x})\, \mathrm{P}_2 x$.

In all like cases we are explaining a certain fact (denoted S_i) or property (denoted P_1) by other facts or properties. In practice, and this goes particularly for pedagogical practice and the questions and answers that crop up in the teaching process, there is also the possibility of encountering cases where the answer to the why-question repeats, or merely expresses in other words what has already been contained in the question, or cases where the property to be explained is explained by a property that is coincident with the first one. [From the logico-semantic point of view this means that the two predicates are synonymous.] In the notation this situation appears as

Question: $(?\vdash)(\vdash P_1 a_1)$.
Answer: $\vdash P'_1 a_1$ and
$\vdash (\forall x)(P'_1 x \equiv P_1 x)$.

These are often cases of verbal explanation which, while offering a certain degree of satisfaction, cannot be deemed satisfactory from the scientific point of view.

Much more complicated are the why-question problem-solving situations where the inferential procedure Inf [---, ...] is not known beforehand. This is particularly the case with various instances of scientific explanation. In science there is also the possibility, moreover, of using not only various forms of deductive inference but also inductive inference. In the case of inductive inference the answering of a question $(?\vdash)(S_i)$ presumes that that in addition to S_{i_1} it is also mandatory to demonstrate $S_{i_2}, S_{i_3}, \ldots, S_{i_n}$ in order for the inductive inference to be supported by statistical generalization, the discovery of stochastic dependences and so on.

To summarize the analysis of the most important types of problem-solving situations we may say that from the point of view of the needs of the methodology of the empirical sciences it is expedient to distinguish three fundamental kinds: those connected with whether-questions, those connected with which-questions, and those connected with why-questions.[24] The characteristic feature of all three kinds of problem-solving

TABLE I

Question	$(?)(S_i)[S_1, S_2, \ldots, S_n, \mathfrak{P}]$	$(?x)(P_i x)[x \in (\hat{x}) P_j x, \mathfrak{P}]$	$(?\vdash)(\vdash S_i)$
Basis of the question	S_i, or as the case may be $(S_1, S_2, \ldots, S_n)$ – sentence	$P_i x$ – sentential function	$\vdash S_i$ – assertion
Uncertainty	Uncertainty in deciding as to the truth value of S_i, or of some other elements of the class $(S_1, S_2, \ldots, S_n)$.	Uncertainty as to the fulfilment of function $P_i x$ by an element of the class $(\hat{x})\, P_j x$.	Uncertainty in the extra-logical reasoning that $\vdash S_i$.
Answer	A decision as to the truth value of S_i, i.e. $\vdash S_i$ or $\vdash \sim S_i$, or the selection of the true elements of the class $(S_1, S_2, \ldots, S_n)$ as the case may be.	A decision as to the element or elements, (all elements, no elements) that fulfil $P_i x$.	Justification of $\vdash S_i$ by means of Inf [---, S_i].

situations is that the pronouncement of the question is the expression of some uncertainty, the true answer being intended to reduce or eliminate the uncertainty. It is also expedient to draw distinctions among what we have called the bases of each type of question. All the major results we have reached may now be expressed in the form of Table I, which indicates the differences in the basis of the questions, in the kind of uncertainty and in the answers.

It should be clear from this table that in each of the three kinds of questions and the associated problem-solving situations the uncertainty that lies behind the question's formulation and is subsequently reduced or eliminated by a true answer is rooted in different aspects: uncertainty as to the truth value of the sentence that forms the basis of a whether-question, uncertainty as to the fulfilment of the function that is the basis of a which-question, and uncertainty as to the extra-logical substantiation of the assertion that is the basis of a why-question.

2. The concepts of 'explanation' and 'scientific explanation'

(a) *The Requirements Imposed on Explanation*

In everyday speech the term 'explain' is used in quite a wide variety of senses and in the most heterogeneous connections. Hence only a part of what we call explanation in everyday speech corresponds to the procedures that we consider to be scientific explanation. It goes without saying that there are not only significant differences between the procedures of explanation in everyday speech and science, but also graduated transitions and certain common features.

We shall deem it appropriate to talk of explanation (leaving aside for the time being the differences, connections and graduated transitions between scientific explanation and explanation generally) if the following circumstances are respected:

(1) Explanation is of the nature of *ontic decision*. This means that explanation relates to the entities of a given universe. Hence the selection of the entities which we shall consider as the objects of explanation depends on the conception of the universe and the scope of the entities which we consider to be its components. According to the diverse goals or connections in which the requirement of explaining sometimes arises,

these objects may be given a closer specification, such as empirically accessible objects, objects with previously determined properties, objects of our subjective experience of the outer world, psychological states and so forth. In view of the logical and semantic nature of explanation, this specification is not, of course, mandatory.

(2) Saying that explanation is of the nature of ontic decision and relates to the entities of a certain universe, we assume that we are in a position to make true assertions about the entities, or at the very least assertions that the objects exist as elements of that universe, that they have certain properties, that they have certain relations to other objects, etc. This means that at the very foundations of the procedure which we have described as explanation lies a certain assertion. From the point of view of the linguistic expression of the requirement of explaining '*something*', this assertion, which is essentially the assertion of this 'something', is usually connected with a why-question.

(3) If the basis of the whole process of explanation is the assertion of what is to be explained, then the process itself consists in the fact that an assertion by which the explanation is made is added to the assertion which is to be explained. This brings us to the two basic components of an explanation:

(a) the *explanandum* (assertion of what is to be explained).

(b) the *explanans* (the assertion by which the explanation is made).

If the assertion S_i is the explanandum, the explanans consists first and foremost of the assertions S_j, S_k, ..., S_n with the possible addition of the assertion of the connection, specified in a certain way, between S_i and S_j, S_k, ..., S_n. This second component of the explanans is generally not given explicitly in the simplest cases, being taken for granted or easily reconstructed from the context. When we say 'the connection specified in a certain way', we mean primarily a connection relating to content, i.e. the semantically relevant connections between S_i and S_j, S_k, ..., S_n. Any fact, event, property or group of properties, decision or process can be explained by giving the causes (causal explanation), purposes, motives, aims (final, or teleological explanation); we explain by means of certain regularities, or in terms of feed-back mechanism, or of the general system of which the phenomenon we are explaining is a component.

(4) In order to consider as an explanation the procedure whereby we give an explanans to the explanandum, we usually require that certain

demands be met, or certain goals be fulfilled. To put it another way, an explanation must satisfy the requirements of the person asking for the explanation. This *pragmatic aspect of explanation* naturally applies mainly to scientific explanation where the demands are much more strictly defined, for example, in that the explanation is to make possible sufficiently reliable decisions, dependable predictions and the like. Nevertheless the pragmatic aspect is of significance to any kind of explanation whatsoever even if what is to be satisfied by it is something purely subjective and not always rational. In as much as we can distinguish between explanation which is satisfactory and explanation which is not, we may naturally also compare various levels of satisfactoriness. This can be seen in the case of phenomena which can be explained in more than one way or by the use of a greater number of explanatory hypotheses.

One characteristic feature of scientific explanation – in contrast to explanation generally – is that some of the circumstances listed here are given quite close specification, and that in addition certain other requirements are met. However, there is no unanimous opinion as to the actual scope of these additional requirements. We shall be discussing some of them in the course of our account at a later stage. For the time being we shall consider those which represent refinements of the four aspects we have already mentioned.

(1a) Since explanation relates to the entities of a given universe, then we speak of scientific explanation when the entities of the universe in question are accessible on the basis of scientifically legitimate methods and procedures. In other words the entities can be directly or indirectly observed, measured, experiments can be made with them, dependable records of them can be used to work with, etc. This further means that we shall not think in terms of scientific explanation for entities that are the objects of fantasy, imagination, belief etc. [This does not of course exclude the possibility of scientifically explaining fantasy, imagination or belief themselves, since these are after all processes that can be observed and objectively investigated. What science will not explain are miracles, or the behaviour and decisions of mythological beings, though it can explain the origin of the ideas of such beings.]

(2a) If we are to explain 'something', the basis of the whole process of explanation is, as we have shown, an assertion, namely the assertion of what is to be explained. Taking the explanandum S_i, then in the case of a scientific explanation it is mandatory that S_i be a truthful, meaningful synthetic statement, and the procedures on the basis of which it can be decided whether S_i is truthful must be scientifically legitimate. If, for example, S_i expresses the results of a description or observation of a certain natural phenomenon, we presume that the results were arrived at (from the point of view of the given level of science) in a reliable manner and have been verified and checked.[25]

(3a) We are willing to consider an explanation as scientific if the explanans operates with such general data as reflect the highest level of cognition attained (the historical aspect) and are capabie of explaining not only the actual phenomena about which the explanandum is a statement but also other like or analogous phenomena or the phenomena of a certain class that has been somehow defined (the nomological aspect of scientific explanation). Furthermore, we are willing to consider an explanation as scientific if the logical relations between explanandum and explanans are conceived in such a way that it is possible, on the basis of the semantic characteristics of the expressions which make up the explanans, to draw conclusions as to the semantic characteristics of the expressions contained in the explanandum.[26]

(4a) The pragmatic aspect of scientific explanation is generally expressed in the demand that the explanation be connected with the possibility of prediction.[27] This demand is of course a very narrow one. A broader and more exact conception of it might be expressed as follows: A scientific explanation may be deemed satisfactory if connected with the possibility of sufficiently dependable prediction, or with that of taking the requisite measures to lead to the fulfilment of the aims intended with the quality intended. (Hence we might formulate Hempel's characterization more generally in the sense that scientific explanation must be potentially predictive or potentially productive.)

(b) *Scientific Explanation and Description*

Scientific explanation must be distinguished from certain other scientific procedures with which it is sometimes confused. What we have particularly in mind are the procedures of interpretation, explication and description. In everyday speech – and even in certain not very precise statements by scientists – the following expressions are not always clearly distinguished: giving a (scientific) explanation of a given fact, giving a description of a given fact, giving an interpretation, and giving an explication. [Use is sometimes even made of such unfortunate expressions as 'giving the interpretation, or explication, of a fact'.]

Since it is mandatory to distinguish ontic and semantic decision, a more exact line should be drawn between procedures having the character of the former and those having that of the latter. The cases of interpretation and explication quite clearly have the nature of semantic decision and hence differ considerably from scientific explanation. Accordingly it is inappropriate to speak of the interpretation or explication of a fact, event or objective process. On the other hand it is fitting to consider the interpretation or explication of expressions, whether they are the results of cognitive activity, expressions of certain conventions, or whatever. What can be interpreted are, for example, formulae in a formal notation, or texts of a normative nature – norms themselves, commands, prohibitions, etc. Explication is likewise of the nature of semantic decision. By explication we understand the procedure of replacing an expression whose meaning is not altogether clear or evident by an expression the meaning of which can be determined with greater exactitude in a language system whose syntactic and semantic rules have been specified more precisely.[28]

A rather more complex matter is the differentiation of scientific explanation and description, since both have the nature of ontic decision. A specific feature of scientific explanation is usually taken to be the circumstance that facts are subsumed under scientific laws;[29] allusions are made to the model of explanation by covering law,[30] and so forth. However, description is also a kind of subsumption in so far as the properties which we ascribe to the object being described are of a general character. If we ascribe to an object of our observation a_1 the property P_1 (from the class of properties $\mathscr{P} = \{P_1, P_2, \ldots, P_n\}$ which are distinguishable in the given context), then the assertion that object a_1 has the property P_1 also means that a_1 is an element of the class created by the predicate of P_1, i.e. $a_1 \in (\hat{x})\ P_1 x$. Thus a description can be called forth by a 'which predicate' question, and the answer which expresses the results of the description will be of the nature of a conjunction:

$$\vdash P_1 a_1 \,.\, P_2 a_1 \,.\, \ldots \,.\, P_n a_1 \,.$$

Each component of the conjunction can of course be replaced by its negation. If the interrelations between the individual elements of the class $\mathscr{P}$ are not known, we have a description in the proper sense of the word. If, however, we uncover any such interdependence and are capable of formulating it in some general form such as

$$(\forall x)\,(P_1 x \rightarrow P_2 x),$$

we may make the transition from description to explanation by uttering the why-question

$$(?\vdash)\,(\vdash P_2 a_1).$$

If a_1 is an element of a non-empty class $\mathscr{X}$ and if $(\forall x)\,(P_1 x \rightarrow P_2 x)$ is the expression of a scientific law, we have passed from the description which states that object a_1 has the property denoted P_2 to a scientific explanation.

Description and explanation are considered to be two of the most important tasks which are performed by means of proper (verbal) signs of the language of science, in particular by means of predicates.[31] However, we sometimes encounter in this connection the simpler view whereby description is made using only a part of the vocabulary of the language of science, namely empirical or observational predicates. Explanation, on the other hand, operates with a different part of the vocabulary, namely 'theoretical' predicates. This simplification can doubtless be upheld historically by the quite justified distinction between the descriptive and theoretical stages of science. Description is doubtless historically and epistemically lower, more elementary and simpler activity than explanation in science, especially with respect to the potentially predictive or productive character of scientific explanation. It is furthermore also true that in description, especially in its simpler forms, empirical predicates may predominate. Nevertheless I contest that the fundamental difference between description and explanation is not in that description operates with empirical predicates while explanation operates with theoretical predicates. It is quite easy to imagine descriptions which would operate with predicates constituted on the basis of a given theory or hypothesis. Indeed the majority of descriptions which we encounter in science are of this kind. And on the other hand it is quite easy to imagine an explanation (for example on the basis of what is called empirical generalization, which we shall be coming to later), which would operate only with empirical predicates.

In description we may operate with both empirical and dispositional predicates; we may operate not only with classificatory but also with comparative or quantitative predicates. It is typical of description that we assign properties to the object being described (descriptum), the result being presented in the form of an expression of conjunctive nature. In description we take no account of the semantically relevant relations between the various predicates. In contrast, explanation is based on the kind of expression of such relations which has a general character with respect to the domain in question. We might, for example, ask why object a_1 has the property denoted P_1 while not at the same time having the property denoted P_2. An explanation might be the discovery that the two properties are incompatible for all objects of the class of which a_1 is an element. Another typical feature of explanation – as opposed to description – is that a major component of what is used to make the explanation is the specification of the connection between the assertion of what we are explaining and the assertion of what is used to make the explanation.

In spite of our having stressed the difference between description and explanation, we should not overlook the fact that in various concrete situations there may be close connections between them. We particularly have in mind here situations where a sufficiently complete and dependable description of many or all the elements of a class under investigation may be an important step towards opening up the possibility of explanation, for instance by discovering general relations which can be formulated in general statements and serve as a component of what we have called the explanans.

(c) *The Most Important Kinds of Explanation*

In our explication of the concept of 'explaining' we have also pointed out the differences and connections between an explanation and a description. As far as the concept of 'scientific explanation' is concerned, we have indicated that in contrast to explanation generally it must meet certain other demands, of which one of the most significant is that it be potentially predictive or potentially productive. However, here too there are certain graded distinctions to be drawn. These can be demonstrated on the basis of the differentiation of verbal explanation, explanation based on empirical generalization, and explanation on the basis of the laws of science.

As an example of *verbal explanation* we might take the famous question and answer in Molière: *Cur opium facit dormire? Quia habet virtutem dormitivam.*[32] The answer obviously does no more than repeat what was contained in the question in other words. From the logical point of view several degrees of verbal explanation can be discerned:

(a) The property denoted P_1 is explained in terms of the property denoted P_1', where P_1 and P_1' are synonymous.[33] If we ask why opium is soporific and answer by saying that it has the property of inducing sleep, it cannot be doubted that we are operating with expressions whose sense in the given context is identical.

(b) The property denoted P_1 is explained in terms of the property denoted Q_1, where though P_1 and Q_1 are not synonymous, Q_1 is a dispositional predicate constituted on the basis of the empirical predicate P_1.[34] It might be asked, for example, why a given substance dissolves in water, with the answer that the given substance is soluble in water. The concepts 'to dissolve in water' and 'to be soluble in water' are not of course synonyms. However, this kind of explanation is still verbal since the dispositional predicate which appears as a major component of the expression comprising the explanans is reducible or definable in terms of the predicates which appear in the explanandum. It is of no great importance whether the constitution of Q_1 is made using reduction procedures or by means of definition; what is significant is that the constitution is based on predicates contained in the explanandum.

(c) The property denoted P_1 is explained in terms of the property denoted Q_1, where Q_1 appears as a theoretical predicate which can be neither defined by means of P_1 nor reduced to P_1. However, it is subsequently discovered that Q_1 is not a theoretical predicate but a predicate that has been constituted *ad hoc*, that may after all be explicated by means of P_1 alone, that arose as the result of a breach of the principle that *nomina non sunt multiplicanda*, and so forth. Examples of this are explanations which operate with concepts such as 'phlogiston', 'fluid', 'thermic substance', 'driving forces', etc. These then are idiosyncratic verbal explanations which, however, from the logical and semantic viewpoint do not even reach the level represented by (b).

All three instances of verbal explanation may, of course, furnish some satisfaction. In case (c), moreover, satisfaction may even be supported by the illusion that the explanation is scientific. From the analysis of the various instances of verbal explanation it is also clear that when we come to construct an explanation which is to be satisfactory scientifically, the problems of the logico-semantic relations of the verbal signs which appear as components of the expressions comprising explanans and explanandum must not be underestimated.

Explanation on the basis of empirical generalization should be distinguished from verbal explanation. The example of this kind of explanation which is usually given in the literature of logical methodology – in the spirit of the principle *exempla non fingo* – is the question and answer: 'Why is this raven black?'. 'Because all ravens are black'.

The characteristic feature of empirical generalization is that for a given universe it fixes certain general relations of the properties designated by empirical predicates. For this reason the explanans, in explanation on the basis of an empirical generalization, contains a general statement. The simplest form of general statement is

$$(\forall)\,(P_1x \rightarrow P_2x),$$

where P_1 and P_2 are empirical predicates.[35]

Another form of empirical generalization is given by Braithwaite (1960; p. 54–55): If A, B, C denote empirical properties, then Braithwaite treats the following statements as empirical generalizations:

> 'Whatever is A and B, is also C, and vice versa';
> 'whatever is B and C, is also A, and vice versa';
> 'whatever is C and A, is also B, and vice versa'.

Using our mode of notation, where the empirical predicates will be P_1, P_2, and P_3, we arrive at

$$(\forall x)\,(P_1x\,.\,P_2x \equiv P_3x),$$
$$(\forall x)\,(P_2x\,.\,P_3x \equiv P_1x),$$
$$(\forall x)\,(P_3x\,.\,P_1x \equiv P_2x).$$

However, for numerous instances of empirical generalization this formulation is too strong, since by way of definition it introduces a further empirical predicate on the basis of a pair of empirical predicates.

The notation of the empirical generalization based on a triplet of empirical precidates, P_1, P_2 and P_3, corresponds to what Braithwaite calls the 'three-factor theory'.[36] If we wished to express our original question in the language of any such 'three-factor theory', it would have to follow the pattern: 'Why is this living creature, which is a raven, black?' And the answer would be: 'Because all living creatures which are ravens are black'. Thus the answer has the form:

$$\vdash (\forall x)\,(P_1x \,.\, P_2x \rightarrow P_3x).$$

Thus it is a much weaker form of empirical generalization than the statement $(\forall x)\,(P_1x \,.\, P_2x \equiv P_3x)$. This form of empirical generalization seems to be of greater weight and frequency than that given by Braithwaite. (It is evident, furthermore, that not all objects that are black are ravens and living creatures, that not all black living creatures are ravens, etc.)

While we would be generally unwilling to consider explanation on the basis of empirical generalization as scientific explanation, it does meet some of the demands of the latter. If we say: If all objects with property A simultaneously have property B, then they will also have property C, then we know on the basis of this assertion that we will find no object with property A, that having property B, will not also have property C. This is to say that even an explanation on the basis of an empirical generalization is potentially predictive or potentially productive. Although the explanans in the case of explanation on the basis of empirical generalization usually does not operate with data which would represent the highest level of cognition attained, it is, provided it is confirmed to an adequate degree, sufficiently reliable to bear out the truth of the assertion lying at the foundations of the explanandum.

Hence it is difficult to make a rigid and unequivocal distinction between explanation on the basis of empirical generalization and scientific explanation. The distinction is justifiable first and foremost historically in accordance with the level of science attained by any given time, and is inevitably influenced by the fact that the interpretation of the attribute "scientific" varies from period to period. This also means that in any actual situation there are numerous transitional stages between scientific explanation – especially when based on scientific laws – and explanation on the basis of empirical generalization. It must also be borne in mind that there has never been a single stereotyped form for what we are willing to consider as scientific explanation, not even for explanation using scientific laws. There is no less problem in setting out a single uniform schema of scientific explanation.

Explanation on the basis of empirical generalization operates only with empirical or possibly dispositional predicates, but scientific explanation may also operate with theoretical predicates. This presupposes that we have the necessary correspondence rules at our disposal.

3. The Typology of Scientific Explanation

(a) *The Commonest Typologies*

In the extensive literature devoted to the problems surrounding scientific explanation we can find quite an extensive collection of various typologies of this kind of explanation. It is not the intention of this volume to add yet another to the list. Instead we wish to point out the need for more exact standpoints by which the various types of scientific explanation can be differentiated. The analysis of these viewpoints will also show that some of the typologies to be found in the literature fail to make the appropriate differentiation.

Hempel, whose works on scientific explanation are customarily, and rightly so, appraised as noteworthy attempts at the analysis of the logico-semantic structure of scientific explanation,[37] distinguishes two basic models of scientific explanation: the deductive-nomological model, and the statistical model. Statistical explanation is further divided into deductive-statistical and inductive-statistical explanation. (For Hempel it is inductive-statistical explanation alone that is the basic form of inductive explanation.) This differentiation of the types of scientific explanation is based primarily on the inferential relations between the expressions which make up the explanandum and explanans respectively. Although Hempel does analyze certain further types within this framework, the differentiation we have outlined remains fundamental for him.

A slightly different typology of scientific explanation has been presented by Nagel (1961), who distinguishes four basic types of explanation: the deductive model, probabilistic explanation, functional or teleological explanation, and genetic explanation. What Nagel calls the deductive model of explanation essentially accords with Hempel's deductive-nomological model. In this model the explanans is constructed as the premises from which the explanandum can be deduced by the respective inferential rules. Naturally there are many complications attached to this model of scientific explanation (which is the basic model for the theoretical natural sciences). The complications relate to the nature of inference, the nature of scientific laws (which have been interpreted in this connection not only as the premises, but also as the inferential rules, the 'instruments', of deduction), and the nature of theoretical predicates and their relations to the empirical predicates which appear in the explanandum. If the relation between the premises which make up the explanans and the assertion expressed in the explanandum is of no more than a probabilistic character, Nagel speaks of probabilistic explanation. So from the logical viewpoint, the premises which make up the explanans in probabilistic explanation do not suffice to guarantee the truth of the assertion expressed by the explanandum. Functional or teleological explanation uses, as components of the explanans, concepts which express the goals, purposes, tasks, functions and other such relations of the object about which the assertion in the explanandum is a statement to a certain system of other objects, to a system of goals, demands, etc. (It is apparent that teleological explanation need not of necessity operate with only such concepts as allow of merely psychological, mentalistic or generally subjective explanation.) Nagel is a little uncertain as to whether genetic explanation is an altogether independent type. It usually operates with a sequence of events, one event being explained by those which precede it in the sequence. Genetic explanation becomes problematical as an indepedent type as soon as it operates with evolutionary laws that have been formulated with absolute precision. In this case genetic explanation would be no more than a special case of the first or second type, i.e. of the deductive model or of probabilistic explanation.

Popper (1934, 1959) makes no great distinction between explanation and causal explanation. Giving a causal explanation of a given event means, in Popper's conception, deducing the statement which describes the event, by means of one or more universal laws together with singular statements as the premises of the deduction. The statements which correspond to what is usually called the explanans[38] include not only general statements which express laws, but also singular statements which express what Popper calls the *Randbedingungen.* From these premises it is then possible to deduce that which corresponds to the explanandum. *Randbedingungen* describe what is usually called the cause, and the conclusion of the entire deductive process describes that which is usually called the effect. Popper's analysis of explanation in essence corresponds to Hempel's deductive-nomological model of explanation.

The fact that the concept of 'scientific explanation' and the various types of explanation are viewed in quite heterogeneous ways can also be illustrated by the standpoint adopted by Braithwaite (1953). This author considers any proper answer to a why-question to be an explanation, the point of the answer being to give some intellectual satisfaction. This intellectual satisfaction may of course be of various kinds and may be presented in a variety of ways. In his analysis of causal explanation Braithwaite draws the distinction between cases where the why-question requires that sufficient conditions must be stated, and those where necessary conditions must be stated for the event about which an assertion is made in the explanandum. As far as the concept of teleological explanation is concerned, Braithwaite's conception does not differ from others'.

(b) *Classificational Criteria of the Typology of Scientific Explanation*

The differentiations of various types of scientific explanation which can be found in the literature sometimes confuse viewpoints and criteria for making the differentiation which are of quite different kinds. Every relatively exact differentiation ought to respect the fact that various types of explanation can be distinguished in accordance with the following different kinds of criteria:

(1) in accordance with the nature of the explanandum;

(2) in accordance with the nature of the laws which appear as components of the explanans;

(3) in accordance with the logical relations between what makes up the explanandum and what comprises the explanans;

(4) in accordance with considerations of a pragmatic nature (particularly the degree of satisfactoriness, completeness, etc., of the explanation).

Let us now take a look in greater detail at the individual types of explanation according to these criteria.

(c) *Classification based on the Nature of the Explanandum*

If we utter a why-question in the sense of a request for an explanation, then according to the *nature of the expressions which make up the explanandum*[39] the following distinct cases, and hence types of scientific explanation, can be distinguished:

(1) We might, for instance, require an explanation of the fact that the square on the hypotenuse equals the sum of the squares on the other two sides, or that the square of the sum of two numbers a and b equals $a^2+2ab+b^2$, or that only one tangent to a circle can be drawn through any one point on the circumference, etc. In these and all similar cases, S_i is of an analytical nature. What is requested is the demonstration of the analytical truthfulness of S_i on the basis of the analytical truthfulness of other expressions and on the basis of the approved inferential rules. In other words, what is required is the proof of S_i. This also amounts to saying that the procedure is in effect one of semantic decision, and hence it cannot be treated as explanation proper.

Although from the formal viewpoint this procedure does have certain features in common with scientific explanation in the domain of the empirical and experimental sciences – which is the domain that is of immediate interest to us here – we shall leave it aside, confining ourselves only to those situations where S_i is a meaningful, synthetic statement.

(2) It is possible to request the explanation of an individual fact or individual concrete event. This may be a socio-historical event, or an individual phenomenon under observation or measurement, an unexpected natural phenomenon, a newly discovered fact or property, a new result in experimental activity, and so forth. If we ask why there was a big drop in temperature today, why there was a mechanical failure in a certain rocket test, what is the cause behind certain complications or difficulties in a patient, we are asking for the explanation of a fact. This means that the explanandum has the nature of a singular statement, which may of course take on a variety of forms; it may, for instance, have the form of a *Protokollsatz*. (Under conditions c at time t an object a_1 was experimentally found to have properties $P_1, P_2, \ldots, P_n$.)

(3) It is possible to request the explanation of a fact that is of a general nature: the fact that ice floats on water (explained by Archimedes' law), the fact that the tone of an approaching source of sonic vibrations changes when the source begins to move away from the observer (explainable by Doppler's principle), etc. In such cases the explanandum can be expressed by means of a general statement, e.g. in the form that all objects that are ice float on water. It then follows that the explanans operates with Archimedes' law.

(4) The explanandum need not always be of a general nature in the sense that it is asserted of all the objects of a class, that has been somehow defined, that they are characterized by certain specific properties. It might be a question of a mass phenomenon to be described statistically: we might ask for an explanation of the fact – as was indeed done by the sociologists of West Germany – that in that country there is a higher percentage of children with a Protestant background than those from Catholic families studying at universities, or more from town families than from country families, more from families living in the more northerly *Länder* than those from the south, more boys than girls, and so on. Here the explanandum has the nature of a certain relation of probability between some of the properties by which the given mass phenomenon is described.[40]

With respect to an explanandum formulated in this way, the explanans may present an analysis of the proceedings, motives, value and prestige scales and other social and psychological factors which have an influence on decision-making, the choice of a way of life, social patterns, etc., of the individual social groups of the population in question. This means that we may find as components of the explanans formulations having the character of either statistical or dynamic laws which express both statistical and deterministic relations.

(d) *Classification Based on the Nature of Laws*

Any ideas as to differentiating types of scientific explanation according to the different natures of individual scientific laws which appear as components of the explanans, ought in the first place to respect two very important circumstances. Firstly, it is a familiar fact that the concept 'scientific law' (in this connection we encounter many expressions which need not always be synonymous with the concept of 'scientific law', e.g. 'laws of science', 'laws of nature') is very vague, unless a more exact explication is given. Formulations which express scientific laws should be distinguished from other general statements, namely from those described as 'lawlike'.[41] The other circumstance is that a uniform classification of scientific laws is not possible, since there are quite a number of different criteria of classification by which to distinguish the various kinds of laws. We shall be leaving aside for the time being the first of these two factors, since discussion of the logical nature of a scientific law, and of the logical, semantic and empirical criteria for distinguishing scientific laws from other similar statements demands more detailed treatment. Discussion of these questions is also closely connected with problems of the instrumentalist and non-instrumentalist conceptions of scientific laws.[42]

The elementary form of a law is generally taken to be a general statement: $(\forall x)(\mathrm{P}x \rightarrow \mathrm{Q}x)$. In the previously defined system $\langle \mathscr{X}, \mathscr{P} \rangle$ this general statement is a law only under certain conditions. The most important of these is that the class $(\hat{x})\,\mathrm{P}x$ must not be empty, i.e. that the statement be a truthful synthetic statement.[43] Taking the predicates P and Q as the designation of properties of certain elements of the class $\mathscr{X}$, then we may consider another (statistical) form of the law to be the sentence: 'The probability that an element having the property denoted as P also has the property Q equals $\mathscr{R}$'. Accepting the convention that this statement might be recorded as

$$\mathscr{P}(\mathrm{Q}, \mathrm{P}) = \mathscr{R},$$

it might seem that the original general statement $(\forall x)(\mathrm{P}x \rightarrow \mathrm{Q}x)$ is logically equivalent to $\mathscr{R}(\mathrm{Q}, \mathrm{P}) = 1$. This is not the case, however. We are entitled to consider that there is a connection between the two forms only when $(\hat{x})\,\mathrm{P}x$ is a finite class and if $\mathscr{R}(\mathrm{Q}, \mathrm{P})$ is interpreted as Carnap's regular *C*-function[44] for finite systems, i.e. if we interpret both $\mathscr{R}(\mathrm{Q}, \mathrm{P})$ and $(\forall x)(\mathrm{P}x \rightarrow \mathrm{Q}x)$ on the basis of the method of state description. From all this it should be apparent that these notational presentations of scientific laws are associated with a number of difficulties caused above all by the uncertainty of their logico-semantic interpretation.

The differentiation of the various kinds of scientific laws, and hence also of scientific explanation, which we are about to describe, pays more regard to practical rather than purely logical considerations.[45] The differentiation is based on the following criteria:

(a) according to the nature of the relations of dependence of facts, objects and properties we distinguish:

(1) laws of the deterministic type,[46]

(2) laws of the statistical type.

The difference which is usually given as characteristic is that laws of the deterministic type enable the accurate and certain prediction of facts, objects and properties, while laws of the statistical type make only probability predictions possible. In reality there is no sharp distinction between the two types; the transition from one to the other can be made possible by a refinement of certain aspects, for example by refining the demands on the exactitude of empirical and experimental discoveries, by extending the domain under investigation through the addition of further entities, and so forth.[47] To return to the last-mentioned notation: it is obvious that if the statement $(\forall x)(Px \rightarrow Qx)$ is a scientific law, it expresses a law of the deterministic type, whereas the statement $\mathcal{P}(Q, P) = \mathcal{R}$ can be considered as a possible expression of a law of the statistical type.

(b) From the point of view of the form of expression of the entities the verbal signs of which appear as components of scientific laws, we may distinguish:

(1) classificatory laws,

(2) comparative laws,

(3) quantitative laws.[48]

If we treat the expressions $(\forall x)(Px \rightarrow Qx)$ or $\mathcal{P}(Q, P) = \mathcal{R}$ as elementary forms of the expression of scientific laws, the differentiation of form depends on whether the predicates P and Q which appear in the formulation of the law, have the character of classificatory, comparative or quantitative predicates. Since the formulation of a scientific law may naturally contain predicates of varying kinds, there can be no clear-cut differentiation of classificatory, comparative and quantitative laws.

Classificatory (or also qualitative) laws operate with predicates which distribute the objects of a given universe into subclasses, e.g. substances into metals and non-metals or conductors and non-conductors of electricity, plants and animals into orders, genera and species, the elements into various groups according to the periodic table, and so forth. Therefore classificatory laws express a very elementary experience (all elements of a given class have such and such properties), which, however, in the course of scientific cognition may be of considerable importance.

Comparative laws operate with predicates of the type sometimes also known as topological or order predicates. These make possible the comparison or ordering of the elements of a given class on the basis of certain previously defined criteria, e.g. according to the relations 'to have a higher temperature', 'to have a greater capacity', 'to last longer'. If we discover that the higher the temperature, the quicker the process of a chemical reaction, then in formulating the discovery we do not yet operate with the quantitative predicates of time and temperature, but only with comparative predicates.

Quantitative laws operate with predicates that are also sometimes known as metrical or numerical. They characterize properties by assigning numerical values to them. The difference between classificatory, comparative and quantitative predicates will be clear from a comparison of the triplets: warm – warmer than – having a temperature of 4 °C; noisy – noisier than – having a loudness of n decibels. Some properties are designated by quantitative predicates in such a way that on the basis of a scale (e.g. that of length, temperature, loudness, voltage, price, mortality, birth-rate, I.Q., etc.) we can determine the relative position on that scale of any object we are describing. The common mode of expression of scientific laws (e.g. $p \times v = RT$, $s = g/2\ t^2$, $f = (m_1 \times n_2)/r^2$, etc.) does not operate with classificatory predicates of temperature, pressure, time, etc., but with the respective quantitative predicates given in units which make up systems of units.

(c) Yet another differentiation of scientific laws can be made on the basis of the role played by the time factor. The formulations of scientific laws (and so, too, of scientific explanation) may be conceived in one of two ways. Either allowance is made for the time factor explicitly in the formulation, or the time factor is left out of account. For this differentiation of scientific laws from the point of view of the time factor, we may barrow J. S. Mill's traditional terminology and distinguish accordingly:

(1) *laws of sequence* and
(2) *laws of coexistence.*[49]

The distinction between laws of sequence and laws of coexistence cannot be identified with that between 'causal laws' and laws of some other type. The question is also involved here whether what we call a causal relation, or one of cause and effect, need of necessity be related to a time sequence of facts in the system being investigated, or whether it should be related to the simultaneous occurrence of two facts. There is no simple answer to this question,[50] since the everyday idea of causality may include both a sequence of states and the simultaneous occurrence of interconnected and interdependent parameters of the given system.

Laws of sequence operate as a rule with the concept of the 'state' of a system. State may be understood as a class of properties (to which there is a corresponding class of predicates) by which a certain object or system is characterized at a certain moment in time. From this it follows that the determination of a state must necessarily involve a time specification. Laws of sequence express the general relations between various states. In the natural sciences they are generally considered to be the most important type of scientific laws since it is they, above all others, that enable us to determine future states, i.e. to make predictions.

However, laws of coexistence, which express the regularity of the simultaneous occurrence of certain characteristics, are also a frequent and no less important type of scientific laws. They are usually conceived in such a way that they express the regular simultaneous presence of certain properties of relations, e.g. that of certain thermic, pressure and capacity characteristics of gases, or of the electrical and thermal conductivity of metals.[51]

(d) The laws of the empirical and experimental sciences may also be classified on the basis of various *levels*, the differentiation of the levels being, of course, only relative and depending on the interaction of the given universe and the 'observer channel' ('experimenter channel', or generally 'channel of science') which we bring to bear in investigating the given universe. Moreover there is always the possibility of reducing the levels to a small number of elementary levels, e.g. the quantum level, the micro-level (molecular level) and the macro-level. The selection of the number and types of levels depends on the nature of the relevant universe as well as on the nature of the measuring and experimental apparatus, and naturally also on certain circumstances of a pragmatic nature.

(e) *Classification Based on the Relations Between Explanandum and Explanans*

Another way of classifying the types of scientific explanation is that based on the relations between what makes up the explanandum and explanans respectively. One of the significant attempts at a logical conception of the relations between explanandum and explanans is the above-mentioned study by Hempel and Oppenheim dating from 1948.[52] Subsequent to the extensive debate which ensued, in which certain objections were raised against the general claims of this conception, Hempel called the schema of explanation whereby the explanandum can be logically deduced from the expressions which make up the explanans, the deductive, or deductive-nomological, model of explanation. This model is therefore only one of the possible models of explanation.

The representation of the relations between what makes up the explanans and what makes up the explanandum by the means furnished by contemporary logic, is naturally not the only possibility. (We shall be indicating some other possibilities in due course.) Essentially, however, a distinction can be drawn between those cases of scientific explanation where we can deduce the explanandum unambiguously and with certainty from the information contained in the explanans, and those where the specification of the explanandum on the basis of the information contained in the explanans is of a probabilistic nature. Viewing the matter in this light we may distinguish two schemata of scientific explanation:

(1) *the deductive-nomological schema*, and
(2) *the statistical schema.*

The principal difference between them lies in the fact that in the first case the explanation is the finding of an explanans which allows of deducing the explanandum unambiguously and with certainty by means of deduction as understood logically, while in the second case it means finding an explanans by which to specify the explanandum with a certain degree of probability. Of course, this is only a very rough distinction. Moreover it must be remembered that this differentiation leaves certain questions unresolved, for instance the question as to how far the rules for deduction are a component of the explanans, or what is the relation between scientific laws and deduction rules, what is the nature of probability statements appearing as components of statistical explanation, and others. (We shall be returning to some of these.)

(f) *Classification from the Viewpoint of Pragmatic Requirements*

There can be quite a number of differentiations of explanations from the point of view of requirements of a pragmatic nature. One of the main matters here is the question of the goals, requirements and demands associated with an explanation. The goals may be both theoretical and practical in nature.

(a) From the point of view of the *theoretical goals* and demands placed on scientific explanations the following types may be distinguished:

(1) causal explanation,
(2) teleological, final or functional explanation,
(3) genetic explanation.

This list is not exhaustive. It should also be realized that the attributes that appear here, i.e. 'causal', 'teleological', 'final', etc., do not always represent the definite or unequivocal description of the respective types of scientific explanation.

The concept 'causal explanation' can be explicated in a variety of ways, depending on the variety of explications of the equivocal and in part vague term 'causality'. Causal explanation need not, of course, be bound to the concepts of 'cause' and 'effect', the less so in that the latter can be given various different interpretations.[53]

As far as the verbal expression of causal relations in science is concerned, it is not the use of particular words that is decisive – as is the case in day-to-day speech. In science it is almost invariably a matter of a stable (in a fixed time and space), reproducible and therefore predictable[54] relation of dependence between certain objects, events or processes. In the light of this, there are indeed grounds for the scepticism expressed by Russell[55] in his refusal to consider the concepts of 'cause' and 'effect' as at all justified in the language of modern science.

Thus we shall not deny the character of causal explanation to cases where the concepts 'cause' and 'effect' do not occur explicitly. This relates in particular to causal explanation operating with scientific laws of the type considered as 'causal'. The very concept 'causal law' has of course a variety of meanings, but at least the following should be considered as necessary attributes of a causal law:

(α) that it be possible to establish (predict, calculate) certain properties of the states of the system under investigation on the basis of previously established initial states and knowledge of the stable relations. This attribute is sometimes also described as the principle of determinism. The fundamental explicit formulation of the principle is the well-known passage from Laplace's *Essai philosophique sur les probabilités* which operates with the idea of an absolute intelligence that is capable, under certain conditions which can never be materialized in practice, of predicting any state in the future. The passage makes absolute what to a lesser degree can be materialized in relatively isolated systems, though the certainty or degree of probability is conditioned by the nature of the investigation and by the level of our knowledge of the system being investigated.

(β) that space be homogeneous and isotropic, and that time be homogeneous. This attribute was formulated by Maxwell. Causal laws express the relationships of states in time and space, but they do so regardless of any particular time or particular space. From this it follows that time-space variables do not figure explicitly in the functions which express the laws in mathematical terms. This attribute also has limitations due to the fact that time and space are not merely the exterior absolutely unvarying background scenery to what goes on in nature but are connected with the actual nature of what goes on. Hence the homogeneity and isotropy are not absolute but merely relative, being limited and valid only under certain conditions.

(γ) the contiguity (of matter, energy or information) of natural events. This contiguity assumes that any procedure of interchange of matter, energy or information is bound by certain temporal and spatial limits, familiar, for instance, from the special theory of relativity. This practically renders the principle *actio in distans*, and all similar ideas, invalid. If we consider causal laws as, for example, the expression of stable, relatively (under certain circumstances) unvarying relations between the states of a certain system, then we are assuming the demonstrable material, energic and informational contiguity of these states due to the fact that (material, informational or energic) transfer and transformation always take place at a finite rate, in a finite space, etc.

Doubts are expressed here and there as to the justifiability of *teleological explanation*, or final or functional explanation as the case may be. These doubts were or are founded in so far as the processes of nature have been explained in terms of 'purposes' or some other such subjectivistically constructed entities which are in effect imputed to those processes by the person giving the explanation. The history of science affords numerous cases of teleological explanation which often merged into theological or verbal explanation.

Leaving aside all the cases where teleological explanation is of an evidently unscientific character, we can still find a complex and heterogeneous spectrum of cases where it is warranted. The main thing to be emphasized is that teleological (or final or functional) explanation relates to the *behaviour* of complex systems. Just as any other scientifically warranted kind of explanation, teleological explanation does offer a certain intellectual satisfaction. It can moreover be divided into two basic types:

The first group includes cases where subjective (i.e. motivational, rational or other psychological) elements participate in determining the behaviour of a complex system. This means that we explain the behaviour (some of the main examples are human behaviour, human decision-making, or the decision-making of social groups as the case may be) of the complex system in terms of goals, motives, requirements, values or any other suchlike factors which may affect the choice of a given decision. Accordingly we might call this kind of teleological explanation *motivational explanation.* It will include not only cases that operate with purely rational motives, but also those with motives that are less rational.[56]

The second group covers cases of teleological explanation which relate to the behaviour of complex systems where there is no direct participation by any subjective elements. (It will be apparent from this that the term 'teleology' is ambiguous.[57]) These are primarily controlled systems, systems with 'goal-seeking' behaviour, systems with autoregulation. Hence the development of cybernetics has meant a great deal in providing additional support for the scientific legitimacy of teleological explanation.[58] So teleological explanation can be applied in the analysis of the behaviour of complex systems with goal-seeking behaviour, whether these be technical, biological or social systems. Systems which regulate temperature in a living organism, the function of leucocytes in the blood, the role of a regulator in an item of technical apparatus, and many other phenomena can also be explained by means of the objective goals or purposes which are to be attained. This mode of scientific explanation does not then invert the direction of real time, as is sometimes objected, but analyzes the task of certain components of a complex controlled system from another point of view.

An important feature of this kind of teleological explanation is the fact that it not only operates with objective categories, but that it is not in conflict with causal explanation. There is also the possibility of converting a teleological explanation into a causal one. A decision in favour of teleological explanation is usually governed by pragmatic considerations. To assess the role of any given component in a complex controlled system, or to assess how far the desired optimum state is attained, it is not always necessary to analyze in detail all the other components.

The connection which teleological explanation has with the concept of goal-seeking behaviour may be also understood as follows: the goal-seeking behaviour of complex systems (technical, biological and social) is so orientated as to preserve some of the parameters of the system and to arrive at desired or (in view of the aims or functions of the system) desirable parameters. The 'mechanisms' intended to maintain the stable temperature of an organism, or certain stable levels in substances, and other homeostatic mechanisms, are fundamentally connected with the *negentropic tendency* which is characteristic of such behaviour. The ordinary schema of causal explanation, which respects the stable relations between states in time and counts with the possibility of future states on the basis of past states that are known, as well as with the homogeneity and isotropy of space, the homogeneity of time and the contiguity of the events of nature, cannot capture this negentropic tendency of behaviour. This confirms the justification of teleological explanation even in cases where it is not necessary to take 'motivational explanation' into account.

The justification of *genetic explanation* in science is disputable. By genetic explanation we generally understand the procedure whereby the explanandum is the description of a certain event in a defined time and space, and the explanans presents a description of other events previous to that which is to be explained; the time sequence and the spatial contiguity of the events are both respected.

Genetic explanation, then, is usually associated with a unique event which is explained by a sequence of other unique events.[59] It is natural, moreover, that any statement about a sequence of events can represent no more than a *selection* from among the totality of events. This selection is what the person giving the explanation considers to be of importance or relevance to the event calling for explanation. Furthermore, it is not without interest to note that the examples commonly given of genetic explanation are taken from history,[60] which means that it is also difficult to reconstruct its logical structure.[61] The enumeration of a mere sequence of unique events corresponds, after all, to the idiographic conception of scientific method in history, and is branded with all the limitations of the method. The familiar motto that the presentation of history is a statement about *wie es eigentlich gewesen ist* allows the possibility of selecting from the sequence of what has been. The original intentions of positivist historiography thus become their very opposite, since the selection must be assumed to be performed according to some sort of scale of values or what has been described in sociology as 'cultural relevance'.

These remarks will have sufficed to indicate the difficulties attached to genetic explanation. It might well be doubted whether it is the most suitable mode of explanation even for the very domain where it is most frequently considered appropriate, namely history. Any of the methodological trends based on the nomothetic conception must by nature give preference to either causal or teleological explanation.

(b) In addition to the differentiation of various kinds of explanation on the basis of theoretical goals, it is also possible, in line with requirements of a pragmatic nature, to consider certain other circumstances, particularly the measure in which an explanation satisfies not only theoretical, but also practical aims.

There is no question that any explanation affords, in varying degrees, the feeling of at least some kind of satisfaction. Having asked a why-question, we may be satisfied with the answer in a greater or lesser degree according to the measure of the (relatively) *a priori* values assumed, explicitly or implicitly, at the moment when we uttered the question. This conception of the measure of satisfaction thus depends on the selection of value systems and scales of values and suchlike. Naturally, this conception may be purely subjective. Nevertheless, there is some point behind the question whether it might not be possible to reconstruct criteria or viewpoints from which to assess the measure of satisfaction, or the measure of dependability of the explanation offered, as the case may be, with greater objectivity. One such criterion might be the *quality* of any decision which is opted for on the basis of the explanation, especially if this quality can be accurately determined or even quantified. This is the case whenever the quality of a decision can be established (*a posteriori*) from the precision of predictions, the average risk rate associated with the decisions, etc. In practice this represents the transition from (relatively) *a priori* criteria to criteria which can be checked, evaluated and qualified *a posteriori*. This transition from *a priori* to *a posteriori* in the evaluation of an explanation assumes, of course, that there is the possibility of basing any assessment on the repetition, regularity and comparison of different decisions. In the case of a scientific explanation this means the possibility of using objective criteria when assessing the measure in which any given explanation is 'potentially predictive' or 'potentially productive'.

4. Scientific laws and their evaluation

(a) *Laws: General Statements or Rules of Inference? A Criticism of Instrumentalism*

Our analysis of why-questions has already served to indicate the significance they have in the explanation of inferential procedures. Wherever an explanation is a procedure in which what comprises the explanandum can be deduced from what comprises the explanans, a considerable role is played by inference. The most elementary schema of explanation is usually taken to be

$$T.C \rightarrow E,$$

where $T.C$ is the explanandum and E the explanans. Hence the explanans is comprised of premises and the explanandum represents the conclusion. This fact is what lies behind the familiar and much debated question as to whether scientific laws represent inferential rules or a kind of qualified general statements which in explanational procedures play the part of premises and so must be true. We find two types of answers to the question. The first viewpoint, which considers scientific laws as inferential rules, is usually described as *instrumentalism*. The second, which treats scientific laws as general statements, carries with it, of course, a number of additional problems connected, above all, with the set of criteria for that which may be considered as a scientific law.

The first hints of the instrumentalist conception of scientific laws go back to Wittgenstein, who provided the background for Schlick's formulation of this standpoint.[62] In the latter's view scientific laws are not statements and cannot be considered as being either true or false; they are expressions that enable us to deduce one set of statements from another set of statements. Ryle gives a similar description of scientific laws.[63] He speaks of laws as 'inferential licences'.

The instrumentalist conception of scientific laws is closely associated with the instrumentalist conception of theoretical predicates in science. The latter are here considered as expressions not having any denotation of their own; they are signs which, if connected with the appropriate syntactic rules, make possible the successful prediction, or other successful systematization or organization, of expressions which do have denotation. Hence the instrumentalist conception of scientific laws, as also the respective conception of theoretical predicates, denies the semantic (and likewise the pragmatic) function of theoretical predicates, admitting their syntactic function only. Instrumentalism is closely related to the positivistic conception of empiricism and to the requirements formulated in the verificational criterion for sense. Scientific laws cannot as a rule be absolutely verified in the spirit of these requirements, i.e. not all possible instances can be scrutinized. Nor does there exist any means of completely reducing theoretical predicates to empirical or 'observational' predicates. Any attempts at satisfying these requirements have of course proved to be in vain, since the very requirements themselves are dubious.

The second viewpoint is much more widespread, though there are quite considerable differences as to how scientific laws and the criteria for their recognition are to be understood.[64] The only point on which there is agreement is that scientific laws are considered as general statements having an undisputed semantic function with respect to the given universe; they can be variously supported by experience, confirmed to varying degrees by the evidence available, tested positively, etc.

Alongside these two conflicting viewpoints there also exist attempts to reconcile them, or at least to make light of the differences between them. One such attempt was that by Nagel.[65] This author points out that the conflict between the realistic and instrumentalist viewpoint is merely one between two different modes of expression. The question as to which is correct is of interest only from the point of view of terminology. Thus for Nagel the acceptance of one or other viewpoint is no more than a matter of selecting the more fitting mode of expression. This conception, which quite openly belittles the contradictions between the two viewpoints, has been somewhat moderated and in part explained by Stegmüller,[66] who emphasizes that the two viewpoints, in so far as they are at all tenable, are of equal value only from the purely logical point of view. (Stegmüller himself adopts the realist standpoint.)

It remains of course true that, from the formal point of view, parallel systems can be constructed in which corresponding expressions can be formulated as rules or as schemata of axioms, both ways leading to comparable results. However, this possibility is confined to formalized systems only, where, in effect, we pay no regard to interpretation. But the problems of explanation in the empirical sciences cannot be limited to no more than the formal, syntactic approach; due respect must also be paid to the semantic aspect. For these reasons the function of scientific laws in scientific explanation cannot be looked on merely as a purely syntactic function.

We shall now attempt to outline in brief some of the more important arguments against the instrumentalist conception of laws in the empirical sciences. These arguments proceed from the assumption that the decidibility of statements in the empirical sciences is itself based on empirical procedures, i.e. observation, measuring, experiments, etc. This means that the statements must also be interpreted, i.e. related to some non-empty universe of objects, their properties and relations, which can be discovered empirically.

The expressions which express scientific laws are used in a scientific explanation in conjunction with empirical singular statements. Taking $T.C \rightarrow E$ as the general schema of scientific explanation, and T as the conjunctive class of scientific laws, i.e. $L_1, L_2, \ldots, L_n$, then the explanans is the conjunction of these scientific laws and the synthetic empirical statement C. From the logical point of view it is obviously scarcely possible to imagine the conjunction of rules and a synthetic empirical statement.

Another objection to the instrumentalist conception of scientific laws arises out of the special situation obtaining in the empirical sciences. It is often a difficult question in the latter to determine which statements are singular and which are of a general nature and are thus able to express scientific laws. Hempel has pointed out in this connection the difficulties connected with the reconstruction in a formal notation of statements formulated in natural language.[67] If we take the example of the statement that our Earth is (approximately) spherical, we may treat it as a singular statement about the individual object called 'Earth'. The selfsame statement may, however, be interpreted as saying that for all points on the Earth's surface it holds that they are (approximately) equidistant from the centre of the Earth. Hempel's objection can be extended further: there is no way of demonstrating any clear-cut or unequivocally definite boundary between expressions which we acknowledge as having the character of scientific laws, and expressions which we consider as empirical generalizations, accidental generalizations, etc. Yet we treat this as a perfectly natural situation and in line with the character of the empirical sciences. From the historical point of view, moreover, we can point to a number of formulations that were, for a time at least, considered to be scientific laws but were not confirmed, or at least not for the whole universe. These problems need not disturb us as long as we consider scientific laws as general synthetic statements related to a certain universe and which therefore cannot be confirmed or tested by means of procedures of the same or similar kind as those used in making decisions about singular synthetic sentences in empirical science. Obviously this coincidence in procedures is incompatible with the conception of scientific laws as inferential rules or licences which make possible the transition from one set of singular statements to another.

Some laws in the empirical sciences are commonly taken as special cases of more general laws; less general formulations can be deduced from more general formulations provided the necessary preconditions are given. For example it is a familiar fact that it is possible to derive, from the formulations which express the relativist laws of motion, non-relativist laws of motion if we take no account of the relation between velocity v and the speed of light c, or in the case where velocity v is negligible with respect to that of light. This means that it is possible to set up the conditions for when a less general law can be deduced from one that is more general. Similarly it is possible to set up the conditions necessary for deducing synthetic singular statements from laws. However, such procedures could not be given any acceptable justification if we were to understand scientific laws as inferential rules.

In more recent times a viewpoint has been expressed by Scriven,[68] which comes fairly close to the instrumentalist conception of scientific laws and against which the same arguments may be raised. Scriven describes scientific laws in scientific explanation as 'role-justifying grounds' of certain sentences in explanation. The background to his ideas is fairly simple and is based on explanation in everyday natural language and on the manner in which facts are explained in history. If we formulate an explanation as the answer to a why-question, then we are explaining a fact, event, etc., by giving what we generally call 'causes'. To the question why ..., we answer because ---, where '...' and '---' are statements which we usually understand as the expressions of cause and effect. In these and similar formulations the explanans contains no expression of any scientific laws. At best there may be a tacit assumption of some connection between causes of a certain kind and effects of a certain kind.

Scriven's viewpoint, based as it is on the elementary forms of explanation occurring in everyday speech, can be broken down by using the formulation of why-questions as described above. The complete answer to the question 'why S_i' is constituted by the assertion of a class of sentences $\{S_j, S_k \ldots, S_n\}$ such that S_i is deducible from $\{S_j, S_k, \ldots, S_n\}$. However, the class of sentences $\{S_j, S_k, \ldots, S_n\}$ contains two different components, which we have expressed by the signs T and C. Scriven is right in the respect that the elementary form of explanation in everyday speech is incomplete since it contains only a certain element or elements of the class $\{S_j, S_k, \ldots, S_n\}$, namely C (usually the expression of causes, an empirically traceable antecedent, etc.). Explanations of this kind are very common in history. In answer to the question why Austria-Hungary declared war on Serbia in 1914 we may be told that it was because there had been an assassination in Sarajevo, namely that of the heir to the throne. The situation is the same when we are told, in answer to the question why it is raining, merely that there has been a decrease in the pressure, or that the sky has been overcast, and so forth. Clearly, the answer gives only C, ignoring T, as well as failing to show how E can be deduced from $T.C$. For these reasons Scriven distinguishes between explanation proper and 'grounds for explanation'.[69]

Reverting to our earlier elementary schema for explanation, we see that for Scriven C alone is the explanation proper with respect to E, while T is the aggregate of the 'grounds for explanation'. Then since T is the conjunctive class of scientific laws, the latter are characterized as 'grounds for explanation', or as the grounds which justify the role of C in the explanation.

The conception whereby in explanation C alone is a legitimate constituent of the explanans with respect to E, in effect turns the relation between scientific explanation and explanation in everyday speech inside out. It is demanded that the higher and more perfect form of explanation, which operates with scientific laws, be judged *sub specie* of simple everyday explanation. However, Scriven goes yet a step further in this conception in pointing out that in numerous cases of the explanation of events in our day-to-day existence, such as in that of explaining that by spilling ink we will stain the carpet,[70] we are simply unable to construct some true universal hypothesis to take on the role played elsewhere by scientific laws.

The assertion that in simple explanation, where we operate with such concepts as 'cause' and 'effect' or which fits the schema 'E because C', it is impossible to construct sentences which would be of the nature of true universal hypotheses, is highly dubious. The very concepts of 'cause' and 'effect' themselves usually presuppose that causes of a certain type are linked with effects of another type for all the objects of a given universe. That is to say that we can construct a general statement to the effect that whenever certain events take place certain others must also take place. The same applies to explanations which adhere to the schema 'E because C'. If we assert that it is raining or will rain because the pump outside is bedewed, or because black clouds have formed, these are naturally not scientific explanations, nor even procedures operating with scientific laws. On the other hand, however, it is apparent that in these and similar cases general statements can be constructed which can then be considered as true universal hypotheses and which, in explanation on the schema 'E because C', we assume to be implicitly valid. The schema of explanation 'E because C', which we use in simple explanations in everyday speech, in fact assumes that we acknowledge the validity of the general statement 'whenever C takes place, E will also take place'. Hence the strict differentiation of 'explanation' (or rather the expressions which constitute the explanans) and 'grounds for explanation', as is demanded by Scriven, is very dubious.

Scriven's assertion as to the incompleteness of simple explanation (i.e. explanation operating with the concepts of 'cause' and 'effect', or explanation on the schema 'E because C') is also only of relative significance. If explanation is 'complete' when in line with the schema $T.C \rightarrow E$, then 'incompleteness' merely lies in the fact that T is not stated explicitly. At the same time, however, as we have already seen, general statements of the type 'whenever C takes place, E will also take place' are implicitly assumed to be valid. Nor can the general statements which can be reconstructed in cases of simple explanation be considered as 'grounds for explanation', but as general true synthetic statements which enable us to decide as to the truth of the statements which constitute the explanandum.

All previous attempts at denying the semantic (and pragmatic) function of scientific laws in the empirical sciences and treating them merely as inferential rules or as 'grounds for explanation' must then be considered as unsuccessful. The function of a law in a scientific explanation, as well as in other scientific procedures, simply cannot be reduced to syntactic relations. Hence this criticism of the instrumentalist conception of scientific laws arrives at similar conclusions as the criticism of the instrumentalist conception of theoretical predicates and the 'theoretical constructs' in the language of the empirical sciences, which likewise cannot be considered merely as syncategorematic words whose semantic function relative to a given sphere of objects can be denied.

(b) *Scientific Laws and Accidental Generalizations*

By not accepting the instrumentalist conception of scientific laws, i.e. by considering them as somehow qualified general statements, there is no avoiding the difficult question of how to separate off those general statements to which we concede the character of scientific laws from those where we are not willing so to do. In other words, can we avail ourselves of a set of criteria capable of enabling us to distinguish between scientific laws and purely accidental generalizations?

For the sake of clarifying the difference between the general statements we consider as scientific laws and those considered as accidental generalizations, let us take a few examples. General statements considered as scientific laws: 'Whenever there is an increase in the temperature of a gas of a certain kind, the volume remaining constant, the pressure of the gas also increases'. 'Whenever a ray is reflected the angle of incidence equals the angle of reflection'. 'Whenever a body falls freely in a vacuum (or if we can ignore the resistance of the medium), the length of the fall is directly proportional to the square of the duration of the fall.' (These examples naturally represent a slight simplification; the majority of scientific laws, especially in the natural sciences, operate with quantitative predicates. However, this does not alter the fact that they are general statements *sui generis*.)

On the other hand we shall consider the following statements and any like them as accidental generalization:[71] 'All the people at present in this room are men'; 'All the trees in this forest are deciduous'; 'All the flowers that I picked in the garden today are roses'.

One very attractive attempt to distinguish the two kinds of general statements and establish criteria appropriate to the difference is that based on 'counterfactual conditionals' and analogous expressions in language. While scientific laws may serve as a basis for deciding counterfactual conditions, accidental generalizations cannot.[72]

The formulation of counterfactual conditions, unlike the ordinary indicative form which is generally used in the expression of general statements, is based on the use of the conditional. We might introduce the two following examples, one each for the two different groups of general statements:

(1) 'If the temperature of the gas increased, the volume remaining constant, there would be an increase in the pressure of the gas.'

(2) 'If it were a flower that I picked in the garden today, then it would be a rose.'

We should be willing to accept formulation (1), provided we know, of course, the respective regularity, i.e. Boyle's law. On the other hand we should scarcely be willing to accept the second formulation (unless of course we had previously scrutinized all possible instances, i.e. all the flowers in the garden, and ascertained that they were roses). In case (1) the basis for decision is knowledge of a scientific law, in case (2) and all similar cases there is no such basis available.

Several aspects should be respected in the analysis of the formulation of counterfactual conditions, particularly the linguistic, logical and cognitive aspects. There are certain problems attached to them which we shall outline in brief.

As far as the linguistic aspect is concerned, it is generally noted that the conditional in this kind of formulation expresses an action that is posited as a possibility and which would occur if a certain condition were fulfilled. In this connection the distinction is sometimes drawn between unreal and real possibility, i.e. possibility understood beforehand not to have taken place, and possibility where this remains open. In a similar sense we distinguish in the methodological literature of logic between unreal conditionals or unreal conditional statements, and subjunctive conditionals or subjunctive conditional statements. If in the present day someone asserts: "If the Persians had won at Thermopylae, the history of Europe would have followed a different course", he is making clear that 'the Persians did not win at Thermopylae'. Unreal conditional sentences of this kind often occur in everyday speech and practice. We might say that if the load of a given lorry were any greater, the provisional bridge would collapse. Or a doctor might state that if the wounded patient had been hospitalized only a few hours later, his life could not have been saved. On the other hand, in the case of subjunctive conditional statements the question is left open as to whether the possibility indicated in the antecedent has or has not taken place.

If we were to apply to counterfactual conditionals, in particular to unreal conditionals, the viewpoints commonly connected in logic with material implication, we should find ourselves in a tricky situation. If we were to consider the formulation of counterfactual conditions as logically equivalent to the ordinary formulation of material implication, we should have to admit that an untrue antecedent implies both the assertion and the denial of the statement in the consequent. But in the case of counterfactual conditionals, especially unreal conditionals, we implicitly assume that the possibility indicated in the antecedent has not occurred, which is to say that we may consider the antecedent to be an untrue statement. The same also applies to the consequent. Thus if we were to treat unreal conditionals as special cases of material implication, we should have to consider them, always and under all circumstances, as true. If we were to consider the antecedent alone as untrue, we should have to accept any consequent, which would mean, for instance, acknowledging as true both the following conditionals:

(3) 'If you picked a large number of roses in the garden, you would prick yourself.'

'If you picked a large number of roses in the garden, you would not prick yourself.'

It is obvious that in these and similar cases we cannot consider the joining of the two sentence components as one where the ordinary viewpoints of truth functions can be applied. Hence the common logical procedures and viewpoints which we apply in the case of extensional combination on the basis of truth functions, are not satisfactory for deciding counterfactual conditions. In order to judge the information contained in the formulation of counterfactual conditions, we must have available extralogical knowledge, or, to put it another way, certain *a priori* information. Therefore, cognitive aspects are of decisive importance for any set of counterfactual conditions.

From the cognitive point of view it is clear that the formulation of counterfactual conditions is usually the expression of a certain generalized knowledge, which is conceived not as the logical but as the *factual connection between unmaterialized but materializable possibilities.* The same connection is expressed in a more condensed form by statements operating with dispositional predicates. Our saying that a certain substance is soluble in water means that we *know* that if it were immersed in water it would dissolve. Thus we do not assume that the possibility is brought about, i.e. that the substance has been immersed in water, yet we do accept the content of the respective formulation of a counterfactual condition, namely that the substance is soluble.

We see then that the fundamental precondition to deciding about counterfactual conditions is some relatively *a priori* generalized knowledge of certain laws or hypotheses which do not require that all possible instances be investigated in order for a decision to be made. On the other hand, this is not the case with accidental generalizations, since there all the possible instances would have to be scrutinized before a decision could be made as to counterfactual conditions. For these reasons – and herein lies the unquestionable intuitive value of Goodman's indication of the role of counterfactual conditions – it is mandatory to distinguish between scientific laws and lawlike statements [73] on the one hand, and statements which we consider as no more than accidental generalizations on the other. Although this criterion for distinguishing lawlike statements and accidental generalizations does represent a kind of intuitive test, it is not capable of furnishing an exact and unambiguous differentiation between the two kinds of statements; nor can it capture the graduations between them.

As we have already pointed out, certain *a priori* knowledge is always of the essence in making decisions about unreal and subjunctive conditional statements. This knowledge not only includes scientific laws or hypotheses but also the knowledge that certain conditions have been realized. One example, given traditionally by numerous authors, is: 'If a match were struck on the side of a matchbox, the match would burst into flame'. Whoever asserts this formulation, understanding it to be true, must assume that the match was not damp, that the inflammatory mixture is not missing from the stick, that the match is struck on the side of the box intended for the purpose, etc. Any assumption of the truth of an unreal or subjective condition might be said to include implicitly the additional assumption that certain relevant conditions have been fulfilled. This requirement of the necessary fulfilment of a given set of conditions is in a way analogous to considerations relating to the stability of the relations between causes of one kind and effects of another, said to apply if and only if the presence of certain conditions is guaranteed. The necessity of determining the set of conditions needed for making a decision as to the conditional 'If there were *A*, there would be *B*' means that alongside the components *A* and *B* there must appear the class of statements *C* expressing the relevant conditions. The combination of statements, or classes of statements as the case may be, *A*, *B* and *C* might seem to be expressible by the means of logical syntax. However, all attempts to do so have proved unsuccessful.[74] Since the relation between the contents expressed by the components *A* and *B* is not of a logical but of a factual nature, it is also impossible to formulate general logical criteria by which to define the properties of the class of statements *C*.

Numerous unsuccessful experiments have also shown that it is likewise impossible to formulate general logical criteria for distinguishing between lawlike statements and accidental generalizations. It is self-evident that not all true general statements can be considered as scientific laws or scientific hypotheses, in spite of their being unquestionably true statements. For example: let us take as a true statement: 'All the objects which were in this urn yesterday were black or white balls'. It is then possible to formulate the obviously absurd conditional (connected moreover with the ostensive determination of the object): 'If this handkerchief were in this urn yesterday it would be either a black or a white ball'. In this and like cases it is clearly superfluous to look for a basis on which to decide the conditional, since it is quite obviously false, false that is on the very basis of the meaning of the proper verbal signs used.

The fact that syntactic criteria do not suffice for distinguishing lawlike statements and accidental generalizations has led to a variety of attempts operating with criteria of a semantic nature. In the context of these, notice is usually drawn to the fact that lawlike statements must bear the marks of *unlimited generality*. However, this requirement is usually not defined positively but negatively: general statements may be conceded to be lawlike if they do not include verbal signs which are in some way or other characterized semantically. These reasons then led to the formulation of various constraints: the unlimited generality of a statement is preserved as long as it does not operate with verbal signs having obviously unique denotation, as long as it does not refer to a limited time-space constellation, as long as it does not operate with proper names, ostensive formulations and the like. The list of constraints may of course be further extended or reduced.

Although these constraints might appear to be sufficiently strong semantic criteria for a general statement to have the character of a lawlike statement, it is not very difficult to show that they are not satisfactory. Let us take that involving the use of proper names. The occurrence of proper names on the one hand and descriptions on the other cannot be a dependable criteria for classifying these statements. Moreover it is well known that the use of proper names can be suppressed by replacing them by descriptions having the same denotation in the given context, e.g. by replacing the proper name 'Napoleon Bonaparte' by the description 'the first emperor of France'. The history of logic tells us of many conceptions which elaborated approaches of this kind in relatively great detail.[75] Similar reservations might be expressed concerning the constraints on the use of ostenstive formulations such as 'the object to which I am pointing', 'the island I am pointing to on the map', and suchlike. If we can express lawlike statements which do not contain proper names or ostensive expressions by equivalent formulations which do contain expressions of this kind, then the occurrence of such expressions in a formulation cannot be considered to be a reliable criterion. Since this possibility is not out of the question, having even been realized time and again in the evolution of science, this criterion falls down as a guarantee of unlimited generality.

This leads to the need to search for other criteria. It might seem that statements which relate to a restricted time-space domain, a finite number of possible instances, or generally to any delimited finite sphere of possible use, cannot be lawlike statements. For these and similar cases it is typical that the formulation of unlimited generality is also conceived negatively or in the form of a constraint. Take the statement: 'The planets of the solar system all revolve around the Sun on an elliptical orbit'. This statement, in spite of its being known as Kepler's law of the motion of the planets, obviously fails to meet the requirement of unlimited generality, since there is – and in the past was – only a finite number of planets known in the solar system. The requirement of unlimited generality is also unfulfilled by any formulation which operates with constants which are valid for only a limited time or space domain, such as the constant for gravitation which is valid for free-falling bodies only close to the Earth's surface. In order to assign the character of scientific law even to those general statements which obviously do not meet the requirement of unlimited generality, and so, too, to satisfy the intuitive conception which has taken root in the evolution of science, some authors[76] have suggested maintaining the criteria we have mentioned, at the same time respecting the usual convention. The example of Kepler's law, or any other similar formulation, represent merely *derived laws* which can be obtained from *basic laws*. Basic laws alone must meet the requirement of unlimited generality.[77]

This means, keeping to the examples given, that Kepler's laws, Galileo's law of free fall with respect to bodies falling free close to the Earth's surface, etc. are derived laws in as much as they can be derived from basic laws, in this case Newton's laws. From the strictly scientific viewpoint, however, this conception, especially as regards the assumption of derivability, is also not entirely watertight: in derivation we have to take into account other circumstances besides what is actually expressed in the basic laws. As Pap (1962; p. 294) has pointed out, Kepler's first law of the motion of the planets is derivable from the laws of Newtonian mechanics on the sole condition that the motion of the planets is determined by solar gravitational attraction alone. This assumption oversimplifies and fails to take account of the mass of the other planets, i.e. it treats their mass as negligible in comparison with that of the Sun. This means that the movements of the planets can be conceived in the framework of the problem of two bodies, in accordance with the principles of Newtonian mechanics.

It is clear, then, that the very procedure of derivation itself is no simple matter, since the laws from which the derivation is made, as well as the statements that are derived, of necessity represent only certain theoretical approximations to the real processes, the actual degree of approximation being perhaps different in different contexts.

Let us take a look at yet another important fact: in order for any derivation to be made, not only do we need certain concrete knowledge relating to a limited time-space domain, but also we must take into account the respective degree of approximation, the demands laid on the accuracy of possible predictions, etc. In practical terms this means that in these and similar situations not only semantic but also pragmatic aspects come into force.

The requirement of unlimited generality is of relevance not merely to statements which refer to individual objects on account of their operating with, for instance, proper names or ostensive formulations. If we were to lay the strictest interpretation on this requirement, then in any statements to which we would concede the character of scientific laws we should have to suppress not only the occurrence of proper names and ostensive formulations but also certain other predicates. We have in mind here predicates of the type 'is French', 'is medieval', 'is Glaswegian', etc., i.e. predicates whose meaning is temporally or spatially restricted. Hence it is recommended that these and any similar predicates – Goodman calls them predicates that are not 'inductively projectible' [78] – be barred from general statements to which we assign the character of scientific laws. The debate as to what predicates can be treated as 'inductively projectible' has not so far led to any positive results. Quite obviously the predicate 'is higher than the Eiffel Tower' [79] is one which is not 'inductively projectible'. Moreover, if we select as the measure of length a given standard, namely the standard platinum-iridium bar stored in Paris, then not even the predicate 'is more than one metre long', or 'is as long as one metre', is 'inductively projectible'. From this point of view any boundary between 'inductively projectible' predicates as permissible in the formulation of scientific laws, and predicates which are not so permissible, is of an entirely conventional nature. This all goes to show that even the criterion of unlimited generality, based on the differentiation of predicates which we have described, is highly problematical. Even the concept of an 'inductively projectible predicate' is problematical, since it assumes the application of predicates to any sphere of objects. It should be said at this point that the familiar illusions of the physicalism of the thirties, namely that any objective phenomenon can be expressed in physicalist language and hence in the terms which the physics of the day employed, have not been confirmed by the subsequent developments in science. If we do accept the thesis of the inductive projectibility of some predicates, i.e. the thesis that these predicates can be used in other spheres, then the projectibility is inevitably to a greater or lesser degree restricted and conditional.

The general statements to which we are willing to concede the character of scientific laws always relate to some non-empty class of objects where there are certain fixed invariant properties or relations. These properties and relations remain invariant within the bounds imposed by certain precisely defined conditions. We might speak of invariance as to time, since, for example, the chemical processes of combination take place in just the same way in the twentieth as in the fifteenth century, or invariance with respect to certain transformations of a physical nature. None of the attempts that have been made to express the properties or limitations of this invariance using syntactic or semantic criteria have proved successful. Thus it appears that criteria by which to ascribe the character of scientific laws to true general statements, or criteria for general statements to be lawlike, are not to be sought merely in syntactic or semantic spheres, i.e. in the domain of the meaning of the expressions used (e.g. in the ban on the use of proper names, ostensive formulations, predicates which are not 'inductively projectible', etc.).

If $(\forall x)(Px \to Qx)$ is a general statement, then the real relation between the properties marked P and Q, which is the relation at the heart of scientific laws, cannot be viewed in terms of all of the properties which belong to material implication.[80] Accordingly, in general statements which have the character of scientific laws the real relation between the properties marked P and Q cannot be mapped by means of material implication. If we choose to call the mapping of this relation *nomological implication*, in accordance with common usage, then nomological implication, as has been shown by Pap,[81] coincides with material implication in abiding by the following rules:

(1) the rule of *modus ponens*;

(2) the rule of the transitivity of implication (also known as the rule of the hypothetical syllogism);

(3) the rule of transposition;

(4) the rule of *modus tollens*.

Unlike material implication, nomological implication does not answer to the rules:

$$(5) \qquad \frac{\sim S_i,\ S_i}{S_j},$$

$$(6) \qquad \frac{S_j}{S_i \to S_j},$$

i.e. the rules corresponding to the paradoxes of material implication, and

$$(7) \qquad \frac{S_i \,.\, S_j \to S_k}{S_i \to (S_j \to S_k)},$$

i.e. the rule of the connection of premises.

However, this differentiation of material implication and nomological implication helps us little, for it can be proved that properties similar to those of nomological implication, namely the fact of not satisfying rules (5)–(7), also pertain to, for example, *L*-implication. Hence this attempt to distinguish material and nomological implication cannot be regarded as satisfactory.

Therefore, in distinguishing between statements having the character of lawlike statements and those which can be regarded as accidental generalizations, syntactic and semantic means alone cannot suffice. This leads to the idea of taking pragmatic devices into consideration. Naturally, the various criteria which might be described as pragmatic may include quite heterogeneous circumstances, especially aspects connected with the degree of inductively understood empirical assertion, the explanatory and predictive value of any given general statement, etc.

One very original attempt to bring together all the above-mentioned criteria in the definition of the concept of 'scientific law' and in distinguishing between general statements and accidental generalizations is a conception to be found in the works of Reichenbach. This author[82] presented his conception of scientific laws in a form which is very original and terminologically not always very lucid. The main emphasis is laid on the fact that 'nomological statements', or strictly speaking 'synthetic nomological statements' (i.e. in Reichenbach's terminology the scientific laws of the empirical sciences or 'laws of nature') must satisfy at the same time a whole complex of conditions. According to Reichenbach nomological statements or nomological formulae are either analytic nomological formulae, in other words logical laws, or synthetic nomological formulae, in other words laws of nature. He accordingly considers the concept 'nomological' as a generalization of the concept 'tautological'.[83]

Reichenbach presents his interpretation of the concept of a 'nomological statement' in the context of his conception of connective operations, where the main ones are taken as conjunction, disjunction, implication, equivalence, etc., i.e. the conjunctions. He then distinguishes between adjunctive and connective operations. The difference between them lies in the ways of reading the truth tables by which the meaning of the operations are usually given. If the truth tables are read one way, i.e. from right to left as presented here, we have the connective interpretation of the truth tables, and analogously connective operations. For the case of the operation expressed by the sign →, see Table II.

TABLE II

S_i	S_j	$S_i \rightarrow S_j$
α	α	α
α	β	β
β	α	α
β	β	α

Hence a connective operation assumes that it is possible to draw conclusions as to the truth values or semantic characteristics of the relevant atomic components on the basis of the truth value, or more generally from the semantic characteristic, of a molecular statement, but not vice versa. In contrast to this, the tables can be read either way in the case of the adjunctive interpretation. The difference between the two interpretations, as indeed Reichenbach duly emphasizes, is of particular importance for implication since adjunctive and connective implication have different meanings. By recording some of the expressions of natural language in a notational form and expressing the various operations merely adjunctively, we seriously oversimplify, which may produce certain difficulties. The difference between adjunctive and connective operations is especially apparent in the context of their empirical interpretation. Adjunctive operations can be verified or falsified only on the basis of the truth values of their atomic components. In this sense adjunctive operations are truth-functional. On the other hand, connective operations are not truth-functional in the same sense. Adjunctive implication and adjunctive operations generally are in no wise capable of mediating information which might be utilizable in prediction. Adjunctive implication is also inadequate to the expression of counterfactual conditions.[84] Reichenbach himself uses the example: 'If snow were not sugar, sugar would be sour'. Although this implication is true in the adjunctive sense, no one would regard it as reasonable. On the other hand we do treat as reasonable the formulation: 'If this piece of metal were heated to a higher temperature, it would expand'. The problem of connective operations is also connected, as Reichenbach has shown, with the analysis of modal concepts such as 'necessary', 'possible', 'impossible', etc.[85]

Although the concept of 'connective operation' is also, for Reichenbach, relevant for the interpretation of improper verbal signs and logical (in Reichenbach 'tautological') implication, he emphasizes that it would be a mistake to reduce the 'necessity of scientific laws' to tautological connective implications. Characteristic for what are usually called natural laws and scientific laws, e.g. physical laws, biological laws, sociological laws, is a synthetic statement based on connective implication.

We have already indicated that only connective operations, and not adjunctive operations, occur in the nomological statements which for

Reichenbach are the legitimate form of expressing what we usually describe as scientific laws. Nomological statements can, however, be divided into two classes: (a) original nomological statements which can be decided directly, (b) those nomological statements which are derivable from those of the first class. Reichenbach calls them derivative nomological statements.

Thus the concept of 'original nomological statement' corresponds to the intuitive concept of 'basic scientific law'. In order to be able to define the concept, Reichenbach introduces a number of requirements which must all be satisfied to give an original nomological statement. The first four requirements are in effect an attempt to give greater precision to the concept of 'unlimited generality'. Statements to which we concede the character of original nomological statements must, according to Reichenbach, be in his words (a) all-statements, (b) general statements, (c) universal statements, and (d) unrestrictedly exhaustive. Although it might appear intuitively that all four requirements express the same thing, Reichenbach does intend them to represent different characteristics.[86]

(a) The concept of 'all-statement' is associated with the occurrence of the general quantifier[87] which relates to the whole scope of the statement. So an all-statement applies to all the objects of a given kind, not merely some of them. Since Reichenbach formulates his criteria of a nomological statement for a formalized language of a certain kind, i.e. for first-order functional calculus[88] with identity, which in his analysis is the object language, this means that statements to which we ascribe the character of nomological statements must begin with the general quantifier. This is not of course to say that in the symbolic notations of nomological statements existential quantifiers cannot occur. Statements to the effect that for all moving bodies there is an extreme velocity which cannot exceed that of light, or that there are certain limits of transmissibility for all communication channels, could be recorded by the appearance of existential quantifiers in the notation side by side with general quantifiers. The occurrence of existential quantifiers in such statements has no effect on the fulfilment of the requirements of an all-statement, which after all corresponds even to the intuitive idea: for all moving objects it holds that if they move, their speed is less than or equal to velocity c.

It should be stressed that the occurrence of the general quantifier is not in itself sufficient guarantee of a statement's being an all-statement. The general quantifier must relate to all the objects of a somehow specified non-empty class of objects. Therefore Reichenbach stipulates that the general quantifier occur in all formulations that are prenex equivalents of the original formulation. So a statement is an all-statement if and only if the general quantifier occurs in every reduced prenex equivalent (of the original statement).

(b) The requirement that nomological statements be general is connected with the foregoing. A statement is general if all its components having the character of statements are all-statements. This means, for instance, that we cannot regard as general a statement which is the conjunction of an all-statement and a singular statement, e.g. 'All metal bodies expand with increasing temperatures and the standard for determining length is a metal body'.

(c) The requirement of 'universality' is linked to the inadmissibility of the occurrence of 'individual-terms' in a nomological statement.[89] By individual-terms Reichenbach understands all terms relating to definite individual objects – proper names, some descriptions and real time-space data. Statements in which individual terms do not occur are called universal statements by Reichenbach. Reichenbach's prohibition also relates to limited predicates such as 'lunar', 'Gothic', 'Parisian', etc. Reichenbach's concept of 'limited predicate' corresponds to the predicates which Goodman considers as not 'inductively projectible'. Thus according to Reichenbach a statement is universal if and only if no individual constants, certain descriptions or limited predicates occur in that statement or any of its prenex reduced equivalents.

The same objections raised against the constraints on the use of proper names, individual descriptions or predicates that are not inductively projectible, can also be applied to this requirement of universality of Reichenbach. This means that the boundaries between statements which we are willing to accept as universal and those where we will not concede this character are far from clear-cut and depend on highly conventional criteria. It is evidently circumstances of a pragmatic character that are and will be decisive in this matter.

(d) A further requirement which must be fulfilled by an original nomological statement is that it be unrestrictedly exhaustive. Intuitively, statements that are not unrestrictedly exhaustive may be described as statements which relate to an obviously limited number of objects – not by virtue of their logical or linguistic form, but by virtue of their meaning. This also applies to statements referring to empty classes of objects. A statement which might be viewed in this way is: 'All the digital computers that were built before the formulation of the mathematical theory of information made use of relays or electronic valves'. This statement is not unrestrictedly exhaustive since we know that before Shannon and Wiener formulated the mathematical theory of information, the number of digital computers with these properties was limited.

A more exact explication of the concept 'unrestrictedly exhaustive' presents of course much greater difficulties than this intuitive interpretation. In the first place we need to know beforehand the number of objects to which a given statement applies. Generally this cannot be decided merely on the basis of the logical or linguistic form of the statement itself, but on the basis of our knowledge. These reasons lead Reichenbach to link the concept of an 'unrestrictedly exhaustive statement' to two signs: a statement is unrestrictedly exhaustive if and only if it does not hold that (a) the statement applies to a finite number of objects, and (b) the fact expressed in (a) is confirmed to a high degree. The lack of clarity in this requirement comes out particularly clearly in circumstances where the limitations placed on the class of objects to which the statement applies are indisputable and hence, to use Reichenbach's words, highly confirmed. Reichenbach's requirement of unrestrictedly exhaustive statements is obviously relativized to the existing level of knowledge and the entire set of facts known to us. This means allowing for the possibility that the further development of our cognition may lead to substantial changes in this relativization.

In addition to these four requirements, which roughly answer the intuitive requirement of 'unlimited generality', Reichenbach introduces another which apparently stems from the view that criteria of a syntactic and semantic nature do not in themselves suffice to guarantee what he himself calls an 'original nomological statement'. An original nomological statement must also be 'demonstrable as true' or 'verifiably true'.[90] This leads to the further assumption that we have available a whole complex of operations which include both deductive and inductive processes and which are capable of deciding about the given statement.

Speaking of inductive procedures, Reichenbach stresses that verification cannot be treated as making absolute decisions, but only practical decisions as to truth. At a later stage he hints that this question is not really a part of the theory of original nomological statements, but comes rather under the 'theory of induction and probability'.[91] This is not to say, of course, that Reichenbach's theory of original nomological statements is somehow incomplete. In his explication of the concept 'verifiably true'[92] he himself stresses the need to respect the possibility of historical changes. If a statement is regarded at a given time as verifiably true and is subsequently shown to be not so, the later decision is the one to be taken as authoritative provided, of course, it is based on more extensive evidence. The previous decision must then be corrected.

The concept of 'demonstrable as true', or 'verifiably true', in the sense in which it is used by Reichenbach is not the same thing as 'truth' as used in the context of singular statements. In the case of original nomological statements we must have at our disposal a sufficiently large class of statements (as a rule expressing the results of measurement, observation, experiment, etc., or the results of other steps in discovery including those of a theoretical nature) to ensure the confirmation of any given nomological statement in a sufficiently high degree. Naturally, these characteristics are all vague. It remains, however, highly debatable whether agreement could ever be reached as to some uniform level for these requirements which would be suited to all the different branches of sciences. Calling the aggregate of these procedures 'nomological confirmation',[93] we can well imagine that the demands as to what nomological confirmation must include will vary from science to science: they will not be the same in physics and geology, or in geology and psychology, etc. These demands will also be historically relative, i.e. conditioned not only by the theoretical level attained but also by the level reached in the evolution of measuring and experimental apparatus, techniques in computation, experimentation, etc. The conclusions which can be drawn from the critical assessment of Reichenbach's conception of scientific laws coincide, therefore, with the conclusions which we expressed earlier in connection with the other conceptions we have been examining. The main conclusions are:

(1) The distinction between scientific laws and accidental generaliza-

tions cannot be set up on the sole basis of criteria of a syntactic or semantic nature.

(2) Alongside criteria of these kinds there also figure criteria of a pragmatic nature which are associated first and foremost with the level of empirical confirmation attainable, the demands placed on this level, and the practical needs which influence the demands.

(3) The dividing-line between the class of general statements which we are willing to consider as scientific laws, and other statements, including accidental generalizations, is not sharp. It is, furthermore, historically relative and historically variable.

The failures in the attempts to find criteria for distinguishing between scientific laws and accidental generalizations in the formal aspects, particularly in circumstances of a syntactic and semantic nature, have naturally served to add weight to the significance of the pragmatic aspects. There are two main sets of criteria here which are of course interconnected and mutually complementary. In a simplified, intuitive form they may be expressed as follows:

(1) The decision as to how far we concede the character of a scientific law to any given general statement depends on the level to which it is confirmed, confirmation involving a whole complex of procedures such as measurement, observation, various kinds of experimentation and the like. Some writers [94] express this by saying that the degree in which statements are lawlike is the degree of their confirmation. Clearly this does nothing to solve the original problem, namely that of the criteria for the differences between scientific laws and accidental generalizations. Responsibility for the solution is merely shifted from one sphere into another.

(2) The decision as to how far we concede the character of a scientific law to any given general statement depends on the ways in which the statement is used and on the quality of the results to which its use leads. Reference is usually made in this connection to the predictive or explanatory power of a scientific law. The predictive power of a general statement lies in our acknowledging it to be a reliable basis for predictions without having scrutinized individual instances directly nor having demanded that such scrutiny be made.

(c) *The Measure of Explanatory and Predictive Power as a Criterion of Qualification of Scientific Laws*

The elaboration of the two criteria just indicated does not solve the original form of the problem, i.e. the problem of the unambiguous (and essentially formal) differentiation of scientific laws and accidental generalizations. Now the conception of the original problem was aimed at deciding how far general statements are or are not lawlike or, in Reichenbach's terminology, how far they have the character of nomological statements. Thus this conception endeavours to solve the problem by taking the general statements as they are and irrespective of their role in the complex of procedures, or complex of tasks generally, that taken in their entirety we call scientific discovery.

On the other hand, the criteria we have alluded to as having a pragmatic nature do respect the conception of science as a complex activity. This further means that neither the measure of confirmation of a scientific law or hypothesis, nor the measure of explanatory and predictive power are taken absolutely but consistently relatively, namely, in the case of the former, with respect to certain available evidence, and in the second case with respect to a certain real or possible event past, present or future.

The problems surrounding the measure of confirmation suggest the extensive analysis of the questions involved in the theory of induction, inductive logic, the statistical theory of experience and the confirmation and testing of hypotheses. One of the pioneering attepts to break the problem down was Carnap's conception of what he called *c*-functions, together with the attempt at a semantic solution of the problem of the measure of confirmation.[95] This experiment did, of course, have serious limitations due largely to the restrictions imposed by the finitistic approach based on state-description. The problems of inductive confirmation are being solved at the present time from various angles, the main trends tending to use the devices of probability rather than those of traditional logic.[96] Since this set of problems exceeds the scope of the present work we shall take a detailed look at only the second complex of criteria, namely those of explanatory and predictive power as criteria of the qualification of scientific laws.

The concept of 'explanatory power'[97] of a general statement was first used by Popper. A remarkable, quantitatively conceived explication of the concept, based on a development of the semantic theory of information, has been given by the group of Finnish logicians headed by Hintikka.[98] The intuitive starting-point of this conception is the view of science as activity which removes uncertainty or ignorance or at least reduces them. Hence, too, scientific laws and hypotheses must carry information which reduces or removes uncertainty. This idea, which is connected with the conception of falsification, was likewise first formulated by Popper.[99] The wealth of content or information in a scientific law grows in accordance with the growth of the class of falsificatory possibilities, i.e. possibilities which a given law does not admit or excludes outright. The larger this class, and so too the lower the level of uncertainty, the higher the informational wealth and explanatory and predictive power of a scientific law or hypothesis. Since the general concept of 'scientific systematization' is sometimes coined for these procedures – as Hempel has suggested – we might consider the systematic power of a scientific law or hypothesis.[100] It should be emphasized that the systematic power of a scientific law or hypothesis is not an absolute category, but one which must be relativized to a certain fact or class of facts.

This relativization is analogous to that which it is profitable to require of semantic information. The fundamental feature of the concept of the 'semantic information' of any given statement is that it is richer in informativeness the greater the number of alternatives it excludes. This means that the information which a statement conveys corresponds to the degree to which uncertainty is eliminated. This further corresponds to the content measure of a statement S_i as introduced by Bar-Hillel and Carnap[101] and definable in the form:

$$\mathrm{cont}\,(S_i) = df\, 1 - p(S_i),$$

where $p(S_i)$ is the measure of probability of what is referred to by the statement S_i.[102] So the content measure clearly grows with the probability of the alternatives which the given statement excludes. Then this content measure can become the basis for the information measure connected with the statement S_i, defined as:

$$\inf\,(S_i) = df\, \log \frac{1}{1 - \mathrm{cont}\,(S_i)} = -\log p(S_1).$$

As Hintikka[103] has pointed out, it is usually not the information furnished by statement S_i in itself that is of interest to us, irrespective of other statements, but rather what he calls *relative information*, i.e. the information afforded by a given statement S_h with respect to what another statement S_e refers to. The definitions given above apply to what might be called 'absolute' information. The simplest form of the relativization of information is presented by Hintikka with the use of the concept of 'transmitted information', or the corresponding 'transmitted content measure'.[104]

In a way similar to that in which Hintikka relativizes the concepts of content measure and (semantic) information, Pietarinen and Tuomela have relativized the concept of 'explanatory (or systematic) power'.[105] The point of departure suggested for the finding of quantitative implements for assessing the explanatory, or systematic, power of a scientific law or hypothesis S_h which is a component of the explanans with respect to the explanandum S_e, is the view which holds that the explanatory or systematic power of S_h is greater as the uncertainty with respect to S_e is reduced. We assume, naturally, that S_h is a general (lawlike) statement and that S_e is neither L-true nor L-false. If $U(S_e)$ denotes the degree of uncertainty attached to S_e,[106] and $U(S_e/S_h)$ the degree of uncertainty attached to S_e provided that S_h holds, it is possible to introduce the measure of explanatory (systematic) power of the statement S_h relativized to the synthetic statement S_e. We denote it as $E(S_h, S_e)$, and it fulfils the following requirements:[107]

$$\text{(R1)} \qquad E(S_h, S_e) = F[U(S_e), U(S_e/S_h)],$$

$$\text{(R2)} \qquad E(S_h, S_e) \gtreqless 0 \quad \text{iff} \quad U(S_e/S_h) \lesseqgtr U(S_e),$$

$$\text{(R3)} \qquad E(S_h, S_e) = \max E \quad \text{iff} \quad U(S_e/S_h) = \min U = 0,$$

$$\text{(R4)} \qquad \max E = 1,$$

$$\text{(R5)} \qquad E(S_h, S_e) \geqslant E(S_k, S_e) \quad \text{iff} \quad U(S_e/S_h) \leqslant U(S_e/S_k).$$

In (R1) F is a real function of two arguments; in (R5) S_h and S_k are two different lawlike statements. On the basis of (R5), F is a monotone decreasing function in its second argument. For the sake of simplicity Pietarinen and Tuomela assume that F is a linear function of the second argument. On the basis of these requirements and the given assumption, the measure of explanatory (systematic) power of the statement S_h is defined with respect to the statement S_e as follows:

$$E(S_h, S_e) = df \frac{U(S_e) - U(S_e/S_h)}{U(S_e)}.$$

In accordance with this measure of explanatory (systematic) power, all statements which reduce the uncertainty with respect to the given statement S_e to the same level, have the same explanatory power. The measure is additive, which means that for two lawlike statements S_h and S_k it holds that

$$E(S_h . S_k, S_e) = E(S_h, S_e) + E(S_k, S_e) \quad \text{iff} \quad U(S_e/S_h) + U(S_e/S_k) - U(S_e/S_h . S_k) = U(S_e).$$

The definition of the concept of explanatory (systematic) power given above stipulates no minimum value for the measure E. One of the possible conditions for the minimum value of E is

$$\text{(I)} \qquad E(S_h, S_e) = \min E \quad \text{iff} \quad U(S_e/S_h) = \max U.$$

On the other hand, however, it is desirable that the maximum measure of uncertainty $U(S_e/S_h)$ be equal to the original uncertainty attached to the statement S_e, i.e. that it be equal to $U(S_e)$. Thus an alternative condition for min E is

$$\text{(II)} \qquad E(S_h, S_e) = \min E \quad \text{iff} \quad U(S_e/S_h) = U(S_e).$$

This second condition corresponds to the intuitive assumption at which we arrive whenever the scientific law or hypothesis is utterly irrelevant with respect to S_e. It should of course be stressed that the two conditions do differ and are mutually incompatible, since they assume different interpretations of the concept of 'uncertainty'. Thus the concept of the 'uncertainty' attached to a given statement is ambiguous and must be given further explication before the measure of explanatory power can be rendered more exactly.

We shall now examine some of the specific conditions which the measure E satisfies. Since the role of S_h is to reduce uncertainty with respect to S_e, it holds that if S_e follows logically from S_j, the latter leaves no uncertainty with respect to S_e, i.e.

$$\text{if} \quad S_h \rightarrow_L S_e, \quad \text{then} \quad E(S_h, S_e) = 1.$$

If we take S_h to be a L-true statement it is obvious that S_h cannot furnish any information with respect to S_e, i.e.

$$\text{if} \quad S_h \text{ is } L\text{-true}, \quad \text{then} \quad E(S_h, S_e) = 0.$$

These conditions are entirely trivial and fully correspond to intuition. It is a much more difficult matter to determine the conditions for the minimum measure of explanatory (systematic) power, i.e. min E. One of the possible solutions would be the situation where the negation of S_e follows logically from S_h. Since S_h L-implies the negation of S_e, the certainty with which we expect that S_e is minimal, and the corresponding relative uncertainty is maximal, i.e.

$$\text{(I)} \qquad \text{if} \quad S_h \underset{L}{\rightarrow} \sim S_e, \quad \text{then} \quad E(S_h, S_e) = \min E.$$

However, yet another solution is possible, operating with the negation of S_h. The measure of explanatory (systematic) power is minimal if S_e follows from the negation of S_h. Hence it holds that

$$\text{(II)} \qquad \text{if} \quad \sim S_h \underset{L}{\rightarrow} S_e, \quad \text{then} \quad E(S_h, S_e) = \min E.$$

Clearly these two alternatives correspond to the alternatives for min E given previously which were based on different interpretations of the concept of 'uncertainty'. The alternative which regards $E(S_h, S_e)$ as minimal when the uncertainty $U(S_e/S_h)$ is maximal and with the provision that $S_h \underset{L}{\rightarrow} \sim S_e$, interprets the measure of uncertainty attached to a given statement as the measure of unexpectedness of what the statement refers to. The second alternative, which treats $E(S_h, S_e)$ as minimal when the uncertainty $U(S_e/S_h)$ equals the uncertainty $U(S_e)$ and with the provision that $\sim S_h \underset{L}{\rightarrow} S_e$, interprets the uncertainty of the given statement as an insufficiency of knowledge of what the given statement refers to.

Both conceptions of uncertainty satisfy the conditions:

$$U(S_e/S_h) \geqslant 0,$$
$$\text{if} \quad S_h \underset{L}{\rightarrow} S_e, \quad \text{then} \quad U(S_e/S_h) = 0,$$
$$\text{if} \quad S_h \text{ is } L\text{-true}, \quad \text{then} \quad U(S_e/S_h) = U(S_e).$$

The intuitive sense of these three conditions is quite clear. In contrast there is another condition which can be conceived in two alternatives which again correspond to the two alternatives for minimum explanatory (systematic) power. By the first:

$$\text{(I)} \qquad \text{if} \quad S_h \underset{L}{\rightarrow} \sim S_e, \quad \text{then} \quad U(S_e/S_h) \text{ is maximal}.$$

By the second:

(II) if $\sim S_h \underset{L}{\rightarrow} S_e$, then $U(S_e/S_h) = U(S_e)$.

Actual instances of the measure of uncertainty are naturally dependent on various circumstances of a pragmatic nature. There is also the possibility of using the above-mentioned content measure (cont) or information measure (inf).

In the first case, if p is the measure of probability of what a given statement refers to, the measure of uncertainty U can be defined thus:

$$U(S_e) = df \text{ cont } (S_e) = 1 - p(S_e).$$

In the second case, based on the measure of information, it holds that

$$U(S_e) = df \inf (S_e) = -\log p(S_e).$$

What we have in these cases is the simple or absolute uncertainty of the statement S_e. The corresponding definitions for the relative uncertainty of S_e with regard to the knowledge of S_h have the form:

$$U(S_e/S_h) = df \text{ cont } (S_e/S_h) = \text{cont} (S_e \,.\, S_h) - \text{cont} (S_h) =$$
$$= \text{cont} (S_h \rightarrow S_e) = p(S_h)\, p(\sim S_e/S_h).$$

Or, to use the measure of information, the form:

$$U(S_e/S_h) = df \inf (S_e/S_h) = \inf (S_e . S_h) - \inf (S_h) =$$
$$= -\log p(S_e/S_h).$$

Since the measure of explanatory (systematic) power of a statement S_h with respect to a statement S_e is defined

$$E(S_h, S_e) = df \frac{U(S_e) - U(S_e/S_h)}{U(S_e)},$$

it is possible to replace in the definiens the measure of uncertainty U by the content measure cont or the information measure inf. In this way we obtain a double specification of explanatory (systematic) power. Pietarinen (1970) develops these possibilities in greater detail, introducing four separate measures for uncertainty and four different corresponding measures of systematic power. The four measures arise out of the combination of alternatives (I) and (II) and the content measure and information measure (cont and inf) for determining uncertainty. In their classical work, Hempel and Oppenheim (1948) used alternative (II) and the content measure cont, thereby obtaining a measure of explanatory (systematic) power having values between 0 (for min E) and 1 (for max E). This measure assumes that $U(S_e) = \text{cont}(S_e)$ and $U(S_e/S_h) = \text{cont}(S_e/S_h)$. Then

$$E(S_h, S_e) =_{df} \frac{\text{cont}(S_e) - \text{cont}(S_e/S_h)}{\text{cont}(S_e)}.$$

Using the probability measure we obtain

$$E(S_h, S_e) = \frac{1 - p(S_h \vee S_e)}{1 - p(S_e)} = p(\sim S_h/\sim S_e)$$

[see also Hempel (1965; p. 287)].

It is clear that the practical application of these and similarly constructed gauges of the explanatory (systematic) power of a scientific law or hypothesis depend on the possibilities of quantifying the respective measure of probability or uncertainty. Different cases seem to call for different ways of determining this measure. For example, in cases where a high degree of structural similarity between the explanatory and predictive procedures can be assumed, it is possible to use as the measure of uncertainty the degree of accuracy of the predictions made possible by the given explanation. If a given explanation is the point of departure for decision-making, i.e. for the selection of various alternative decisions, the specification of uncertainty can be based on the qualities which these decisions are discovered to possess. (The very concept of the 'qualities of a decision' can itself be explicated in various ways of course. One possible explication is that based on the concept of 'average risk'. One decision out of a given set of possible decisions is of a proportionately better quality, the greater the degree to which it reduces the risk-level associated with it.)

The practical application of these gauges may relate to tasks of the most diverse kinds. We shall mention here three typical tasks from which their significance will be apparent.

(a) If we have a certain set of lawlike statements $\{S_{h_1}, S_{h_2}, \ldots, S_{h_n}\}$ which appear to be possible systematizational devices with respect to S_e, e.g. as possible components of a potential explanans with respect to S_e, as a basis for prediction, or as a point of departure for the selection of a decision, it is profitable to compare the explanatory, or systematic, power of each of the separate elements of the class $\{S_{h_1}, S_{h_2}, \ldots, S_{h_n}\}$. If $E(S_{h_i}, S_e) > E(S_{h_j}, S_e)$, we will evidently give preference to S_{h_i}.

(b) Since the measure of explanatory (systematic) power is additive, it is better, in the circumstances that

$$E(S_{h_i} . S_{h_j}, S_e) > E(S_{h_i}, S_e),$$

or

$$E(S_{h_i} . S_{h_j}, S_e) > E(S_{h_j}, S_e),$$

to operate with both S_{h_i} and S_{h_j}.

The need to use both statements ceases to apply if, for instance, one of the two statements S_{h_i} and S_{h_j} is irrelevant with respect to S_e, being therefore in no position to raise the measure of explanatory power of the other. Referring again to the second alternative interpretation for the measure E, S_{h_j} is irrelevant with respect to S_e if

$$U(S_e/S_h) = U(S_e).$$

In that case it holds that

$$E(S_{h_i} . S_{h_j}, S_e) = E(S_{h_i}, S_e) + E(S_{h_j}, S_e) =$$

$$= \frac{U(S_e) - U(S_e/S_{h_i})}{U(S_e)} + \frac{U(S_e) - U(S_e)}{U(S_e)} = E(S_{h_i}, S_e).$$

Thus we are justified in replacing $\{S_{h_i}, S_{h_j}\}$ by the first of them alone.

(c) The problems of the reduction and constitution of the elements of the class $\{S_{h_1}, S_{h_2}, \ldots, S_{h_n}\}$ are encountered in yet another typical task. It is often desirable to find a minimal conjunctive subclass of elements of class $\{S_{h_1}, S_{h_2}, \ldots, S_{h_n}\}$ such as ensures max E and which, with respect to S_e and the whole class $\{S_{h_1}, S_{h_2}, \ldots, S_{h_n}\}$, represents the most suitable general component of a potential explanans for S_e. There is, of course, always the possibility that this subclass will be empty.

One very important task in practice is to find what we might call *sufficient explanatory (systematic) power* with respect to S_e. Let us assume that the possibilities we have of carrying out explanatory or predictional (or generally systematic) procedures are limited. The claims made on the accuracy of the results of these procedures may also be limited. The concept of 'sufficient explanatory (systematic) power' can be introduced with some greater precision on the basis of the following intuitive considerations: the explanatory (systematic) power of S_h with respect to S_e is greater in proportion to the reduction of uncertainty with respect to S_e if S_h is known. (If $U(S_e/S_h) = U(S_e)$, the explanatory power is minimal, i.e. zero.) Thus the difference between the values of $U(S_e)$ and $U(S_e/S_h)$ may be taken as the starting-point. Taking the sign $\mathscr{E}$ to denote the maximum admissible uncertainty, which can be interpreted as the maximum level of admissible risk, the admissible level of inexactness in prediction, etc., it must hold that

$$U(S_e) - U(S_e/S_h) \geqslant \mathscr{E}.$$

The concept of 'sufficient explanatory (systematic) power', which we denote as AE, can then be defined thus:

$$E(S_h, S_e) \text{ is } AE(S_h, S_e) \text{ iff } U(S_e) - U(S_e/S_h) \geqslant \mathscr{E} \quad (\mathscr{E} > 0).$$

Let us assume as an example that we are to predict the coordinates of a moving body. The prediction could be made on the basis of classical dynamics, but also, and with greater accuracy, on the basis of relativistic dynamics. The important thing is whether the differences between the two procedures, or to be more exact the differences between the results they lead to, are at all material, and also whether they do not affect the admissible level of inexactness. Since, metaphorically speaking, we do not use a slide-rule where the multiplication tables suffice, nor a digital computer for what the slide-rule will tell us, we can generally be content with sufficient explanatory (systematic) power. This further means that out of the class $\{S_{h_1}, S_{h_2}, \ldots, S_{h_n}\}$ which may appear as the general component of a potential explanans for S_e we may omit all elements that are irrelevant for S_e, but also additional elements whose omission does not lower the overall uncertainty below the level $\mathscr{E}$. In other words:

$E(S_{h_i} . S_{h_j} . \ldots . S_{h_n}, S_e)$ is the sufficient explanatory (systematic) power iff $U(S_e) - U(S_e/S_{h_i} . S_{h_j} . \ldots . S_{h_n}) \geqslant \mathscr{E}$.

Having analyzed the concept of 'explanatory (systematic) power' we can now return briefly to the problems involved in the differentiation of accidental generalizations and lawlike statements. The attempts that have been made to find syntactic or semantic criteria for the purpose have proved unsuccessful. Part of the reason for the failure of these attempts is due to the fact that they took the respective statements in isolation, ignoring their role in certain given procedures and their relations to other statements. The introduction of the concept of 'explanatory (systematic) power' does not, of course, solve the problem of the criteria for lawlike statements in the form in which it was originally framed. It does, however, make it possible to relativize the problem. Of course, it still does not lay open any universally valid sharp dividing-line between scientific laws and accidental generalizations, though it does allow of evaluating the relevance of general statements with respect to a given, usually singular, empirical statement.

5. Scientific explanation and decision-making

(a) *Deductive Models of Scientific Explanation*

A number of authors writing in the twenties and thirties have already pointed out the connection between scientific explanation and deductive procedures. The basis here was that in scientific explanation the usual procedure is to subsume singular statements relating to facts or events that are to be explained, under general statements expressing what we call scientific laws. The conditions which warrant the subsumption must always be stated; for they too have the nature of singular statements.[108] One of the first works to contain a relatively more exact formulation of this conception of scientific explanation is Popper's *Logik der Forschung*[109] with its explication of the concept of 'causal explanation': giving a 'causal explanation' of an event means deriving a statement which describes that event, or fact, from statements expressing scientific laws or hypotheses and statements expressing specific conditions. Thus scientific explanation, which is in this sense an inferential procedure, calls for nomological statements on the one hand, and singular statements expressing the specific conditions on the other hand.

In this way an explanation is treated as the justification of the statement which comprises the explanandum by means of other statements. These other statements, i.e. the explanans, fall into two components: statements expressing the specific conditions, antecedence or other necessary conditions, and statements which express the scientific laws or hypotheses that are competent with respect to the given explanandum. (*Note:* The term 'antecedence' is more in line with Popper's conception which starts from the analysis of causal explanation and the assumption of operation with 'laws of sequence'. However, it is possible to operate in an analogous way with laws having the character of 'structural laws' or 'laws of coexistence' which do not give the immediate expression of the causal connection of states, e.g. Ohm's law, Boyle's law and others. Hence, the term 'antecedence' is limited in meaning with respect to causal explanation.)

This view of the structure of scientific explanation has been expressed by Hempel and Oppenheim[110] in the following schema:

$$\begin{array}{ll} \textit{explanans:} & C_1, C_2, \ldots, C_m \\ & \underline{L_1, L_2, \ldots, L_n} \\ \textit{explanandum:} & E \end{array}$$

All the signs in this schema are assumed to refer to statements, $C_1, C_2, \ldots, C_m$ being statements expressing specific conditions, and $L_1, L_2, \ldots, L_n$ statements expressing scientific laws.

This schema assumes that the statements which comprise the explanans have the character of premises, the explanandum being the conclusion which follows from them. Hempel and Oppenheim also formulated the conditions of the adequacy of an explanation, i.e. conditions to guarantee that the statements which make up the explanans are capable of explaining what is expressed in the explanandum. There are four such conditions:

(R1) The explanandum must follow logically from the explanans.

(R2) The explanans must contain the relevant scientific laws essential to the deduction of the explanandum (or at least statements which follow from scientific laws).

(R3) The content of the explanans must be empirical.

(R4) The statements that comprise the explanans must be true.

The formulation of these four conditions is considerably restrictive. As Hempel later acknowledged, they relate to no more than the simplest cases of deductive-nomological explanation in the empirical sciences. Considerable restriction is imposed by, for instance, the requirement of the truthfulness of the statements which make up the explanans. The history of science tells us that such statements may be highly confirmed on the basis of available evidence and so regarded as true, yet must be modified or even abandoned with the discovery of new data which refute them. This is what led Hempel to distinguish later between true explanation, which meets all the requirements, and potential explanation, which meets them all except that of the truthfulness of the statements in the explanans.[111] Since there can never be any absolute guarantee for the statements that make up the explanans, nor is there any absolute way of ensuring that any given explanation is true or merely potential. Since it is never possible to establish more than a certain measure of confirmation of these statements, the actual measure being always relativized with respect to the level of knowledge obtaining at any given time, science usually operates with only potential explanation, the reliability of which is dependent on the measure of confirmation.

The concept of 'potential explanation' was introduced by Hempel and Oppenheim as an analogy to lawlike statements, i.e. statements having all the formal appurtenances of scientific laws with the exception of truth. If we introduce for the two components of the explanans the signs T (i.e. the conjunction of all $L_1, L_2, \ldots, L_n$; this conjunctive class must not be empty) and C (i.e. the conjunction of all $C_1, C_2, \ldots, C_m$), then the ordered pair $\langle T, C \rangle$ comprises the potential explanans with respect to E, if the following conditions are met:

(1) T is composed of essentially generalized statements, i.e. statements in which the quantifiers are always at the beginning and which are not equivalent to any singular statement.[112]

(2) E is deducible from T and C in the given language.

(3) T is compatible with at least one class of atomic statements the consequence of which is C but not E.

It will be obvious that the aim of these conditions is to eliminate circular explanations or apparent explanations; this means that the explanans must operate with information which is not contained in the explanandum. Lengthy critical discussion of these conditions has shown that even they are not sufficient to exclude completely circular or apparent explanations. Many authors have therefore tried to formulate further conditions and so too more stringent criteria for the recognition of the ordered pair $\langle T, C \rangle$ as a potential explanans with respect to E.[113] There is of course no overlooking the fact that neither these conditions nor any extension of them, in so far as they relate to circumstances of a syntactic or semantic nature, are capable of capturing the real cognitive quality of a given explanans, since they take no account of the explanatory or predictive power of the general statements with which the explanans operates.

The deductive structure of this schema of scientific explanation has to ensure the transfer of those semantic characteristics which relate to the statements that comprise the explanans, to the statements which make up the explanandum. Since the explanandum is naturally assumed to be composed of a true statement, strictly speaking the assertion of a true statement, we need to formulate requirement (R4). We have already shown that there are serious difficulties in applying that requirement consistently wherever there is no possibility of making a complete verification owing to our having only a certain satisfactory degree of confirmation to go by. Then of course we are asserting to be true a conclusion deduced from something that is only highly confirmed.

(b) *Statistical Models of Scientific Explanation*

The deductive schema of scientific explanation described above, which treats scientific laws as one of the premises of a deductive procedure, assumes that the scientific laws which figure as components of an explanans have the form of nomological statements. These are the reasons that lie behind the occasional use of the designation 'the deductive-nomological model' of scientific explanation.

Scientific practice, however, since the nineteenth century, has introduced statistically formulated laws which are based on a different principle from nomological statements. Although statistical laws have – in the spirit of Laplace's classical conception of probability – been regarded as a kind of provisional arrangement which is the expression of our imperfect knowledge and which will be replaced by laws of the deterministic type when the state of our knowledge improves, it must be emphasized that the significance of statistical laws in modern science, far from waning, has in fact increased. It was especially the developments in physics, which long served as the prototype of deterministic science (in Laplace's sense), and then developments in thermodynamics and quantum theory that showed the crucial significance of statistical laws in science which cannot be regarded subjectivistically, i.e. as the product of our insufficient knowledge of nature, but as the expression of the objective relations and processes in nature, and which simply cannot be reduced to laws of the deterministic type.

Similarly as laws of the deterministic type, expressed in the form of nomological statements, are required to function as components of explanans with respect to explananda which relate to individual facts or events, statistical laws are also expected to be able to explain a given fact, event or phenomenon. In this case, however, there is a radical difference between the role of statistical laws (using the above schema) and laws having the character of nomological statements. For this reason the deductive schema, where the explanandum is the conclusion that follows from the relevant premises, cannot be used. A limited possibility for using it does however remain provided that the explanandum is not a singular statement but a derived statistical law. Hempel speaks in this connection of *deductive-statistical explanation.* Let us assume, for example, that we are to explain why on average *m* boys are born in Prague as against *n* girls. The explanation of this phenomenon, which is not of course a singular event, but a certain statistical regularity, can be made quite legitimately by reference to the fact that in the whole of Czechoslovakia, or in the population of Europe as a whole, the ratio of boys to girls born is m/n. In fact what we have here is a derivation of a less general statistical regularity from a more general one. Only in a case like this can a schema be constructed which would be analogous to the deductive nomological model of explanation.

The difficulties of using this model if we operate with statistical laws in the explanans, having a singular statement in the explanandum, become apparent from an analysis of the 'statistical syllogism'. Let us assume that all the objects in the class $(\hat{x})\mathrm{P}x$ will very probably (with a probability greater than or equal to a value not very different from 1) have a property Q. If there were complete inclusion of the classes $(\hat{x})\mathrm{P}x$ and $(\hat{x})\mathrm{Q}x$, it would hold that if an object a_1 had the property P, it would also have property Q, hence:

$$\frac{\begin{array}{l}(\forall x)\,(\mathrm{P}x \rightarrow \mathrm{Q}x)\\ \mathrm{P}a_1\end{array}}{\mathrm{Q}a_1}$$

Some authors regard the kind of inference usually described as the statistical syllogism to be justified in practice:[114] Let a_1 be an element of the class $(\hat{x})Px$ of which it holds that almost all its elements have the property Q. Then the following appears to be justified:

Nearly all objects with property P also have property Q.
Object a_1 has property P

Object a_1 almost certainly has property Q

The formulation of this inference can of course be given various modifications. The proportion of objects with property P and those having Q as well might, for example, be specified by a fraction, not very different from 1; or it might be required that the conditional probability $p(Q, P)$ be greater than or equal to some value that has been previously established. It should be further noted that expressions such as 'certainly', 'almost certainly', 'very probably', which occur in the formulation of the statistical syllogism or other analogous formulations, may be interpreted as 'epistemic modalities'.

Hempel has shown [115] that these statistical syllogisms cannot be taken as a logical basis for scientific explanation, since they are procedures (Toulmin speaks in this context of quasi-syllogisms) which lead to conflicting conclusions. This can be seen if, for instance, we assume to be valid the premises of the inference mentioned above, i.e. the statements: 'Nearly all objects with property P also have property Q' and 'Object a_1 has property P'. Let us further assume that there exists a non-empty class $(\hat{x})P'x$ of which it is known that almost all its elements do not have property Q. If then the intersection of classes $(\hat{x})Px$ and $(\hat{x})P'x$, while being slight, is non-empty, then it is possible that a_1 comes under this intersection. However, if we acknowledge the above statistical syllogism as acceptable, we must also admit the following statistical syllogism:

Almost all objects with property P′ do not have property Q
Object a_1 has property P′

Object a_1 almost certainly does not have property Q.

The conclusions of the two procedures are clearly at variance. Thus we must reject the statistical syllogism as a reliable logical basis for scientific explanation. Hempel speaks in this connection of the ambiguity of statistical explanation.[116] What is meant by ambiguity follows from the next two inferences whose conclusions are at variance:

Inference 1:
More than 99% of the people living in Paris are not Moslems
Salam lives in Paris

It is almost certain that Salam is not a Moslem.
Inference 2:
More than 99% of Arabs are Moslems
Salam is an Arab

It is almost certain that Salam is a Moslem.

There are no doubts that in both cases the statements which correspond to premises from which a conclusion is derived, are true. In neither case, however, do we have a deductive inference, so that the source of error lies in the mode of deducing the conclusions from the true premises. Since it is not a matter of deducing a less general statistical regularity from a more general one, the problem here is one of inductive statistical explanation and cannot be solved on the basis of these schemata. This is not of course to say that statistical laws and hypotheses are not of service as a device in explanation or as a basis for prediction. It is obvious, however, that inductive statistical explanation is of a fundamentally different character from that of the above mentioned deductive nomological model of scientific explanation. The same also applies to prediction on the basis of statistical laws, which, in practice, is equally important as prediction on the basis of laws of the deterministic type. This is true, for example, in the case of predictions in the realm of genetics, the use of data on the half-life of radioactive substances, etc.

The characteristic feature of inductive statistical procedures is the fact that they do not guarantee the transfer of semantic characteristics, or epistemic modalities as the case may be, from the premises to the conclusion, i.e., in the case of a scientific explanation, from the statements that constitute the explanans to the statements that make up the explanandum, in the same way as in cases of deductive explanation if the explanans has the character of true or verified assertions. So it is possible to speak of true deductive nomological explanation (in Hempel's sense). However, as Hempel has himself recognized, there is no usable analogous model in the case of inductive statistical explanation.[117]

If it were possible to construct a usable analogy, we might construct a schema of inference similar to that of deductive nomological explanation. If we consider $(\forall x)(Px \rightarrow Qx)$ as a nomological statement to which corresponds the statistical law $p(Q, P) = r$ (where $0 \geqslant r \geqslant 1$), it should be borne in mind that the role of $(\forall x)(Px \rightarrow Qx)$ with respect to an individual fact, e.g. the fact that Pa_1, differs fundamentally from the role of $p(Q, P) = r$. Particularly if we interpret the concept of 'probability' in the spirit of the frequency conception we are faced with the problem of the justification of using the term as applied to a single fact or event.[118] In addition to the interpretation in the spirit of the frequency conception there is also the possibility of using the conception of logical probability. A logically probable statement is one which makes it possible to *estimate* the value of the given statement in the light of evidence found.[119] (There do exist other conceptions of probability and inductive inference based on certain psychologizing concepts, various conceptions of subjective or psychological probability, conceptions of probability based on the concept of rational belief, and so forth. The detailed analysis of these and other conceptions of probability and probability inference exceeds the scope of this work.)

The deductive nomological model of scientific explanation would correspond to the following schema of inductive statistical explanation:

explanans: $p(Q, P) = r$ ($r \geqslant \varepsilon$, where ε is a number not very different from 1)
Pa_1
$\sim\sim\sim\sim\sim\sim\sim\sim$ with probability r:

explanandum: Qa_1

The wavy line here indicates that in effect this is not an inference in the true sense of the word.[120] It is easily shown that this in no way removes the ambiguity of statistical explanation which in form is quite analogous to the cases of the statistical syllogism which we have demonstrated above. Let us imagine that $a_1 \in (\hat{x})\, Px$ and simultaneously $a_1 \in (\hat{x})\, P'x$, P and P′ being different from each other. If, however, the intersection of the classes $(\hat{x})\, Px$ and $(\hat{x})\, P'x$ is non-zero, there are two ways of proceeding that cannot be placed out of the question:

$p(Q, P) = r \quad (r \geqslant \varepsilon)$
Pa_1
$\sim\sim\sim\sim\sim\sim\sim\sim$ with a probability of r
Qa_1

and simultaneously

$$\begin{array}{l} p(\sim Q, P') = r' \quad (r' \geqslant \varepsilon) \\ P'a_1 \\ \sim\!\sim\!\sim\!\sim\!\sim\!\sim\!\sim\!\sim \text{ with a probability of } r' \\ \sim Qa_1 \end{array}$$

Since both r and r' are relatively high probability values, we get here two contradictory conclusions to both of which we can ascribe a relatively high probability. Thus it is indeed necessary to find a solution to get round this difficulty.

One such solution offers itself in the shape of Carnap's requirement of total evidence. This requirement runs: "In the application of inductive logic to a given knowledge situation, the total evidence available must be taken as basis for determining the degree of confirmation".[121] This means taking into account absolutely all the knowledge that is available in the given situation and at the given instant. To return to the last example, this means respecting simultaneously that $a_1 \in (\hat{x})\, Px$ and $a_1 \in (\hat{x})\, P'x$. Thus the statements $p(Q, P) = r$ and $p(\sim Q, P') = r'$ cannot be taken as a starting point in isolation but in the form

$$p(Q, P.P') = r^*,$$

or, as the case may be, we might take the statement

$$p(Q, P.P'') = r^{**}.$$

If we know, for example, that $a_1 \in (\hat{x})\, p_1 x$, $a_1 \in (\hat{x})\, P_2 x, \ldots, a_1 \in (\hat{x})\, P_n x$, and this represents the entirety of our knowledge in a given situation and at a given moment, then

$$p(Q, P_1 . P_2 \ldots . P_n) = r,$$ [122]

will be a component of the explanans.

There are a number of objections that can be raised against this application of Carnap's requirement of total available evidence. Firstly, it is debatable how far the application of this requirement, which demands that all available knowledge be taken into account at a given moment and in a given situation, allows us to remain in the domain of inductive inference. Carnap himself of course emphasized that the requirement of total available evidence is not a principle of inductive logic, being rather one of methodology aimed at facilitating the application of inductive logic.

Another objection is that the application of the requirement increases without profit the number of dependences with which we must calculate, no account being taken of their respective relevance. Yet the varying relevance of the elements of the class $\{P_1, P_2, \ldots, P_n\}$ to the dependence between Q and P_i must be respected. Carnap[123] – in the spirit of the conception suggested by Keynes – has specified the concept of 'relevance'. If we consider the pair of elements P_i and P_j of the class $\{P_1, P_2, \ldots, P_n\}$, then:

(a) P_j is *positively relevant* to Q on the basis of P_i iff $p(Q, P_i . P_j) >$
$> p(Q, P_i)$.

(b) P_j is *negatively relevant* to Q on the basis of P_i iff $p(Q, P_i . P_j) <$
$< p(Q, P_i)$.

(c) P_j is *relevant* to Q on the basis of P_i iff P_j is either positively or negatively relevant.

(d) P_j is *irrelevant* to Q on the basis of P_i iff either

(1) $p(Q, P_i . P_j) = p(Q, P_i)$; or

(2) $(\hat{x})\, P_i x \cap (\hat{x})\, P_j x = 0$.[124]

For interpreting the concepts of 'relevance' and 'irrelevance', the possibility of using a graphic schema, corresponding to Carnap's graphic interpretation of the relations between deductive and inductive logic, presents itself. If we formulate the dependence of predicates on the basis of the dependence of the classes $(\hat{x})\, Qx$, $(\hat{x})\, P_i x$ and $(\hat{x})\, P_j x$, then the irrelevance of P_j to Q on the basis of P_i might be expressed as in Figure 6.

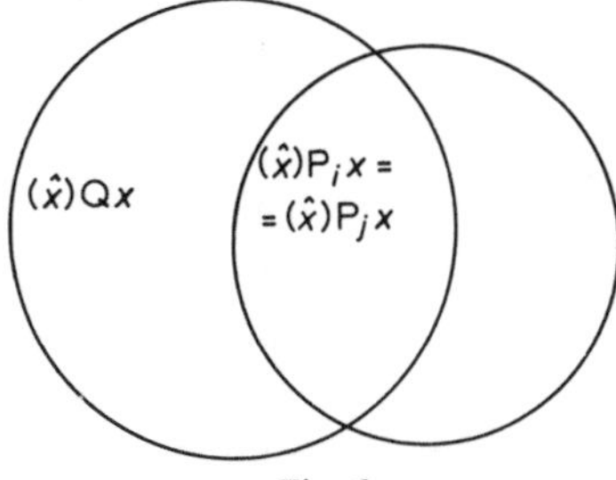

Fig. 6.

In the given case P_j is irrelevant to Q on the basis of P_i because $(\hat{x})\, P_i x =$
$= (\hat{x})\, P_j x$, so that $p(Q, P_i . P_j) = p(Q, P_i)$. P_j is likewise irrelevant to Q

on the basis of P_i if the intersection of $(\hat{x})\ P_ix$ and $(\hat{x})\ P_jx$ is empty, which can be expressed schematically as in Figure 7.

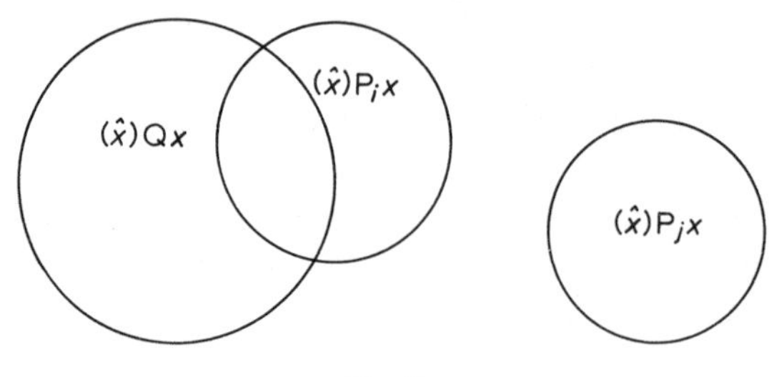

Fig. 7.

(In this Figure the intersection of $(\hat{x})\ Qx$ and $(\hat{x})\ P_jx$ is also empty, which is however only one of the various possibilities. It can equally well be imagined that the intersection of classes $(\hat{x})\ Qx$ and $(\hat{x})\ P_jx$ be non-empty, or that the class $(\hat{x})\ P_jx$ is a subclass of $(\hat{x})\ Qx$ or that $(\hat{x})\ P_jx$ is the complement of class $(\hat{x})\ P_ix$.)

The requirement of total available evidence can thus be moderated by respecting the relevance of the elements of the class $\{P_1, P_2, \ldots, P_n\}$ and by excluding those elements that are irrelevant to Q on the basis of a certain P_i. Irrelevance is of course relative and may change with the replacement of P_i by another element of the class $\{P_1, P_2, \ldots, P_n\}$.

In conceiving a statistical hypothesis of the type $p(Q, P_i)=r$, where $P_i \in \{P_1, P_2, \ldots, P_n\}$, there is profit to be gained in taken account of other elements of that class. This procedure may also lead to an increase in the value of r, possibly by so much that r equals one. In that case it becomes possible to speak of the determinization of a statistical hypothesis.[125]

Against the application of Carnap's requirement of total available evidence in inductive statistical explanation it must be objected that its consistent application must lead to highly complex and tedious procedures which are generally beyond the bounds of real possibility, whether in explanation or in prediction. The procedures might indeed be reduced if we had at our disposal the requisite data as to the varying relevance of all the available parameters and were in a position to eliminate the irrelevant parameters. But this may call for yet other knowledge and so too for the extension of the procedures necessary. Moreover, if we were to demand the extension of the knowledge necessary to the practical determinization of the respective statistical hypothesis, this would amount to the denial of the justification of inductive statistical explanation. In other words, to treat the requirement of total available evidence as the requirement of finding hidden parameters would mean subordinating inductive statistical procedures to deductive procedures and so to a 'deductive conception' of any kind of induction.

Another attempt to solve the problem of replacing the requirement of total available evidence has been undertaken by Hempel. Taking again the basic form of inductive statistical explanation as a starting-point:

$$\text{(IS)} \qquad \begin{array}{l} p(\mathrm{Q}, \mathrm{P}) = r \quad (r \geqslant \varepsilon), \\ \mathrm{P}a_1 \\ \rule{3cm}{0.4pt} \quad \text{with probability } r, \\ \mathrm{Q}a_1 \end{array}$$

it is possible to require that $(\hat{x})\,\mathrm{P}x$ be the smallest class containing the element a_1. However, Hempel, in his first work on statistical explanation,[126] pointed out the difficulties attached to this possibility. He was led to require that inductive statistical explanation – unlike deductive nomological explanation – be always relativized to a certain 'knowledge situation'. This knowledge situation can be treated as the class $\mathscr{K}$ of statements regarded as true in the situation in which the inductive statistical explanation (or corresponding prediction) is made. Hence, as Hempel stresses, we cannot speak of inductive statistical explanation as such, but of inductive statistical explanation with respect to $\mathscr{K}$. Thus we are faced with formulating a requirement which, to put it briefly, would enable us to decide between two statistical hypotheses which lead to contradictory conclusions. Let the two hypotheses be in the form:

$$\begin{array}{l} p(\mathrm{Q}, \mathrm{P}) = r \quad (r \geqslant \varepsilon), \\ p(\sim\mathrm{Q}, \mathrm{P}') = r \quad (r' \geqslant \varepsilon), \end{array}$$

it being simultaneously valid that $a_1 \in (\hat{x})\,\mathrm{P}x$ and $a_1 \in (\hat{x})\,\mathrm{P}'x$. Then we are to choose between $p(\mathrm{Q}, \mathrm{P}) = r$ and $p(\sim\mathrm{Q}, \mathrm{P}') = r'$. Hempel's attempt was to formulate the *requirement of maximal specificity*[127] to make the choice possible. Let us imagine a scientific explanation in the form (IS). If $\mathscr{I}_\alpha$ is the conjunction of the two components of the explanans in (IS), and $\mathscr{I}_\beta$ is the conjunction of all the elements in $\mathscr{K}$, then (IS) is rationally acceptable with respect to $\mathscr{K}$, provided the following condition is fulfilled: if it follows from the conjunction of $\mathscr{I}_\alpha . \mathscr{I}_\beta$ that $a_1 \in (\hat{x})\, P^*x$ and $(\hat{x})\, P^*x \subset (\hat{x})\, Px$, then the statistical hypothesis which specifies the conditional probability of Q with respect to P* must also follow from the conjunction $\mathscr{I}_\alpha . \mathscr{I}_\beta$, i.e. $p(Q, P^*) = r^*$. Then $r^* = r$ or $p(Q, P^*) = r^*$ is a theorem of the mathematical theory of probability.

Hempel assumed that the requirement of maximal specificity was capable of eliminating one of the two rival statistical hypotheses, i.e. of deciding in favour of one of the two following procedures:

$$\begin{array}{ll} p(Q, P) = r \quad (r \geqslant \varepsilon) & \\ Pa_1 & \\ \sim\!\sim\!\sim\!\sim\!\sim\!\sim & \text{with probability } r, \\ Qa_1 & \end{array}$$

$$\begin{array}{ll} p(\sim Q, P') = r' \quad (r' \geqslant \varepsilon) & \\ P'a_1 & \\ \sim\!\sim\!\sim\!\sim\!\sim\!\sim & \text{with probability } r', \\ \sim Qa_1 & \end{array}$$

For if $\mathscr{K}$ contained components from the explanans of both procedures, both r and r' representing a high probability, and if both procedures fulfilled the requirement of maximal specificity, it would follow from $\mathscr{K}$ that

$$p(Q, P.P') = p(Q, P) = r,$$
$$p(\sim Q, P.P') = p(Q, P') = r',$$
$$p(Q, P.P') + p(\sim Q, P.P') = 1,$$

and so

$$r + r' = 1.$$

However, this becomes impossible if it holds that $r \geqslant \varepsilon$ and $r' \geqslant \varepsilon$ and ε represents a high probability. This means then that only one of the procedures meets the requirement of maximal specificity.

It is true that Hempel's requirement of maximal specificity succeeds in eliminating one of the two rival statistical hypotheses, yet it offers no guarantee that the elimination is the one that is appropriate. Attention has been drawn to this by Wójćicki[128] in his criticism of Hempel's conception of inductive statistical explanation. Another of the critics of Hempel's conception, Humphreys, has shown that the requirement of maximal specificity is so restrictive that it practically precludes any statistical hypotheses.[129] Similarly other authors have also produced counter-examples the aim of which has been to prove that the requirement of maximal specificity is either too liberal or too restrictive. It was on the basis of these criticisms that Hempel subsequently modified the original version of his requirement of maximal specificity. In order to prevent the inappropriate elimination of one of two rival statistical hypotheses, it is required that a scientific explanation in the form (IS) operate with a probabilistic statement having the character of a statistical lawlike statement. Hempel laid down two requirements which must be met for the statement $p(\mathrm{Q}, \mathrm{P})=r$ to be acknowledged as a statistical lawlike statement:[130]

(N1) In a statement of the shape $p(\mathrm{Q}, \mathrm{P})=r$ neither of the predicates Q and P may have, for purely logical reasons, a finite extension. The requirement that predicates having finite extension be excluded is to be understood as meaning: A predicate Q has finite extension for purely logical reasons if, for example, the statement $(\forall x)\,(\mathrm{P}x \rightarrow \mathrm{Q}x)$ is *L*-equivalent to the conjunction $\mathrm{Q}a_1 . \mathrm{Q}a_2 . \mathrm{Q}a_3$. In other words, Q denotes a property which pertains to a finite number of objects, so that the given general statement is *L*-equivalent to a finite number of singular statements. This requirement represents a certain analogy with the requirements given earlier whose fulfilment was meant to ensure the character of a nomological statement (or lawlike statement).

(N2) No complete sentence composed of a predicate, which occurs in a statement of the form $p(\mathrm{Q}, \mathrm{R})=r$, and a proper name (undefined) is *L*-true.

By a complete sentence composed of a predicate and a proper name Hempel understands the sentence which arises out of the use of that predicate in relation to the proper name in the sentence. This requirement is also analogous to those which lay behind the attempt at a conception of the semantic criteria of 'unlimited generality' of lawlike statements. Disregarding the very problematical nature itself of Hempel's concept of 'statistical lawlike statements', it can be shown in the case of each of these requirements that they are far from adequate to the unambiguous distinction of statements of the form $p(\mathrm{Q}, \mathrm{P})=r$ – which we shall regard as statistical laws, from those statements where we do not admit this characterization.

For the further modification of the requirement of maximal specificity Hempel introduces certain other concepts, namely 'maximally specific predicate', and 'stronger predicate': predicate P_1 is stronger than predicate' P_2 if $(\forall x)(P_1x \rightarrow P_2x)$ is L-true. A predicate M is the maximally specific predicate with respect to Qa_1 in $\mathscr{K}$, if (1) M is L-equivalent to the conjunction of the predicates that are relevant to Qa_1 in $\mathscr{K}$; (2) M does not logically imply either Q or $\sim$Q; (3) there is no other predicate stronger than M which meets conditions (1) and (2).

Hempel's modified requirement of maximal specificity then runs: A procedure of the form

$$\begin{array}{ll} p(\mathrm{Q}, \mathrm{P}) = r \quad (r \geqslant \varepsilon) & \\ \mathrm{P}a_1 & \\ \sim\!\sim\!\sim\!\sim\!\sim\!\sim\!\sim\!\sim\!\sim & \text{with probability } r, \\ \mathrm{Q}a_1 & \end{array}$$

can be described as an inductive statistical explanation with respect to $\mathscr{K}$ if the following conditions are met: Let the predicate M be either the maximally specific predicate with respect to Qa_1 in $\mathscr{K}$, or a stronger predicate than P while statistically relevant to Qa_1 in $\mathscr{K}$. Then $\mathscr{K}$ contains the statistical lawlike statement $p(Q, M) = r$, where $r = p(Q, P)$.

It will be evident that Hempel's considerably complicated attempt at establishing criteria for distinguishing a rationally acceptable model of inductive statistical explanation has its weak points as well as certain difficulties that are no less that those attached to the application of Carnap's original requirement of total available evidence. We have already shown that we cannot accept as satisfactory the attempt at a semantic differentiation of statistical statements which are lawlike and those which, if true, we acknowledge as statistical laws or statistical hypotheses. It must further be stressed that the practical determination of what Hempel calls a maximally specific predicate may be fraught with considerable difficulties.

Let us now note yet another difficulty associated with inductive statistical explanation, namely that connected with the fact that we are in reality operating with epistemic modalities. The procedure that Hempel described as inductive statistical explanation in fact operates with various different epistemic modalities. Expressions such as 'is highly confirmed that...', 'is highly probable (with probability $r \geqslant \varepsilon$) that...', 'is certain (true, verified, etc.) that...' must be treated as different epistemic modalities.[131] These epistemic modalities can be differentiated according to the degree of certainty or probability with which something is asserted. This means, for instance, that 'it is certain (true, verified, etc.) that...' is a stronger epistemic modality than 'it is almost certain that...', 'it is highly probable that...' and suchlike.

Hempel endeavours to construct schemata of inductive statistical explanation on the basis of analogy with the deductive nomological schema with its premises and conclusion. However, this does not obviate the question of how to justify the derivation of a statement bound by stronger epistemic modality from statements, some of which are bound by weaker epistemic modalities.

One way round these difficulties has been suggested by Stegmüller.[132] Expressions of the kind 'it is practically certain that...', 'it is highly probable that...' and the like, which are what we operate with in schemata of inductive statistical explanation, while they can be treated as epistemic modalities, are, in this sense, incompatible with how these epistemic modalities can be related on the one hand to the statements that make up the explanans, and on the other hand to the explanandum. Stegmüller infers from this that these expressions have the character of relational concepts and relate to what he describes as the relation between premises and conclusion in the schema of scientific explanation. Since it is debatable whether what we have in the schema of inductive statistical explanation are really premises and conclusion in the proper sense of the word, it might be said that these expressions relate to the procedure that is reconstructed between explanans and explanandum. However, it seems easier to accept the account which relates epistemic modalities not only to the statements that make up the explanans but also to this procedure. This means then that all the expressions to which may be ascribed the character of epistemic modalities in inductive statistical explanation, are used in practice in two meanings or two different levels: as expressions which qualify the statements which constitute the explanans, and as expressions which qualify the nature of the procedure which we reconstruct between explanans and explanandum.

We can now conclude that what Hempel and certain other authors call inductive statistical explanation fails to meet the demands and expectations placed on scientific explanation. This is also why what constitutes the explanans in the case of inductive statistical explanation does not really have the character of premises from which the explanandum can be deduced. These are the reasons that lead Stegmüller to conclude his extensive analysis of inductive statistical explanation [133] by inferring that there is really no such thing as inductive statistical explanation, but only inductive statistical *Begründung*.

It is natural that the statistical character of scientific hypotheses and laws does not mean that they are without cognitive value. The fact that we are not able, on the basis of the laws of half-life, to explain why one particle disintegrated while another did not (strictly speaking we are not able to explain in the original sense of the term 'explain') does not at all mean that we cannot calculate on this statistical law or make successful decisions on the basis of it. Statistical laws and hypotheses do also reduce our uncertainty with regard to a given situation and so we are justified in speaking of their systematic power. This particularly applies to procedures where laws and hypotheses of this kind are the point of departure for predictions, the selection of an optimal decision from a set of possible decisions, etc.

(c) *The Decision Model*

Hempel's schema of scientific explanation, especially the deductive nomological model of scientific explanation, is based on the assumption that the explanandum can be *inferred* from the data contained in the explanans. Attempts at reconstructing inductive statistical explanation, or inductive statistical *Begründung*, are also based on similar assumptions. From this point of view it is possible to speak of *inferential models of scientific procedures*. [134]

Starting from the most elementary schema of explanation, as given above and recorded in the form

$$T.C \rightarrow E,$$

where $T.C$ is the explanans and E the explanandum, it is easily demonstrated that a similar schema is used in the analysis of other scientific procedures. This makes it possible to speak of the isomorphism of a certain group of scientific procedures. Structural similarity can be shown to exist between Hempel's deductive nomological model and the logical schema of diagnosis elaborated by Ledley and Lusted.[135] The diagnostic procedure described by Ledley and Lusted breaks down into three basic components:

(1) The first component which the doctor (or computer) must have at his disposal is a description of the patient's state, usually carried out on the basis of certain symptoms or a complex of symptoms. This state description might be denoted *Sd*, and be understood as a statement as to the presence or absence of the respective symptoms. (In the simplest instance this statement can be imagined as the conjunction of atomic statements as to the presence or absence of the symptoms as determined individually.)

(2) The next thing that must be available is what Ledley and Lusted described as 'medical knowledge'. Medical knowledge can be understood as statements as to the connections between individual symptoms and ailments. In diagnostic procedure the doctor must always take into account the aggregate F of medical knowledge, which can be imagined as the conjunction of statements as to all established connections between symptoms and ailments. Medical knowledge may also be understood in the sense that F makes it possible to ascribe certain complexes of ailments to certain complexes of symptoms.

(3) The actual diagnosis is then the statement D as to what ailment may be inferred.

The procedure which leads to the *inference* of the diagnosis might be expressed as follows: If the doctor has his disposal certain medical knowledge (F) and if a certain state description (Sd) is found in the patient, the relevant diagnosis can be inferred from F and Sd. This may be recorded in the form:

$$F.Sd \rightarrow D.$$

Ledley and Lusted regarded this schematic record (which they write in the form $F \rightarrow (Sd \rightarrow D))$ as the fundamental formula in diagnosis. However, since the relationships between symptoms and ailments are not of a deterministic, but of a probabilistic nature, Ledley and Lusted suggested a certain probabilization of the conception, based on the concept of conditional probability and Bayes' well-known theorem for testing hypotheses.

It is true that a causal interpretation can be ascribed only to the deductive nomological model. This means that the circumstances given in the explanans may also be described as causes. On the other hand, what we call symptoms are not generally interpreted in terms of cause and effect and causal relations. For these reasons some authors deem it mandatory to distinguish between causal antecedence and causes, and causal explanation and reasoning operating with symptoms and statements about symptoms.[136] Although the difference between causal concepts and symptoms unquestionably has certain ontological connotations, there is clearly in either case the possibility of selecting the optimal solution, provided, of course, we do have the necessary knowledge at our disposal. Thus for the analysis of explanatory (and hence in part also predictional) procedures and diagnostic procedures, isomorphic models can be used.

There are three fundamental kinds of data to be distinguished in any of these scientific procedures:

(a) initial data of an empirical character, such as data on the patient's state in diagnostic procedures, data as to causal antecedence in the case of causal explanation or prediction, the data that make up the empirical component of the explanans in the deductive nomological model of scientific explanation and prediction, etc.

(b) resultant data, such as the establishment of a diagnosis, data as to causal sequence, the parameters of the future state of a system, etc.

(c) data on the relations between the initial and resultant data, usually conceived as data on their general interrelations together with the respective transformational procedures.

Inferential models of scientific procedures of the kind we have been speaking about are based on the assumption that it is possible to deduce the data in (b) on the basis of the data in (a) and (c). From this it is also clear that the concept of 'relation' is here of paramount importance, as well as the nature of the relation and the level of our knowledge about it.

Taking as our basis procedures in which we are to establish the data in (b) on the basis of the data in (a) and (c), we might also say that in all such cases what we are dealing with is the transfer and processing of data and hence a communication process in which can be discerned input and output space as well as the communication channel, including certain transformational procedures. Inferential models of scientific procedures (all such procedures being operations with data) further assume the possibility of deducing data in the input space on the basis of data in the output space and on the basis of knowledge of the functioning of the communication channel including the respective transformational procedures.

The communicational presentation of procedures of these kinds has certain advantages: It is not based on the priority of the deterministic functioning of the channel and so assumes a broader spectrum of relations between the initial and resultant data, i.e. between the elements of the input and output space (or vice versa). In addition, the communicational mapping implicitly allows for the possibility of noise in the communication channel. Thus the essential matter is to *decide* as to elements of the input space (or output space) on the basis of available knowledge about the elements of the output space (or input space respectively) and knowledge as to how the communication channel functions. This leads us to consider *decision models of scientific procedures*, e.g. of scientific explanation, diagnosis, etc.

From the intuitive point of view, decision-making differs from inference particularly in that in the former the emphasis is on the weighing of individual alternative decisions and the results of the activity connected with the respective decisions, and on the search for optimal decision functions, i.e. rules to govern the selection of the most appropriate decision from the set of possible decisions. Viewed in this way, inferential models, especially simple deductive nomological models, are in effect special cases of the more general situation of mapping in decision models.

Decisions models also have the advantage of capturing more adequately certain of the special features of the procedures we have in mind, namely features connected with the necessity of respecting the requirements of methodological finitism, as well as certain basic tendencies which are usually associated with these procedures. One of the main tendencies is that of *optimizing* a given procedure, for instance by reducing the uncertainty involved or the risk connected with any particular decision, refining predictions of future states, finding more exact, more adequate and more complete explanations, achieving more dependable diagnoses, and so forth. In other words the matter in hand is to meet the demands laid on the given procedure in a higher degree than hitherto. In addition to the optimizing tendency there is that of *economizing* the given procedure. By this we mean that when the procedure comes to be performed we do not exceed certain limits, e.g. demands on time and space, on the amount and quality of the means required, the amount and quality of the data needed, etc. We might imagine the refinement of predictions by the inclusion of other additional data in our determination of future states. If, however, this leads to such complications and delays that the required results are not reached in time, any such extension of data ceases to have sense. Imagine the analogous situation in diagnosis. The two tendencies (i.e. of optimization and economization) cannot be applied in isolation, but only in some kind of interrelation. This relates particularly to the necessity of extending, or admissibility of reducing, the data with which we calculate in any given procedure. If the demands laid on the quality of the solution, e.g. on the accuracy of predictions, the adequacy of a diagnosis, etc., are not met, it becomes necessary to try to optimize the decision processes, by, for example, extending or refining the data in the output space, refining the decision functions with which we operate, etc. Furthermore it should also borne in mind that we never have available more than limited potentials, limited means, delays, capacities, etc.

Decision models of scientific procedures of the kind mentioned are based on the assumption that these procedures are operations with data *sui generis*, the transfer and processing of information. From the intuitive viewpoint it is clear that the greater the wealth of information afforded by the output space, the higher the quality of decision-making relating to the evaluation of the elements of the input space on the basis of the elements available at the output space. It also holds that the more adequate the decision process proper (or, to put it another way, the better our knowledge of the relevant relations), the higher the level of the results attained. The possibility of assessing the adequacy and varying degree of relevance of the means by which the operation is performed offers itself here. Equally important are the limits to the possibility of decision-making, i.e. the *problems of decidability*. If we look on the given procedures as processes of the transfer and processing of data, then these problems are analogous to the familiar problems of transmissibility in the mathematical theory of information. If the channel which we have at our disposal is lacking in properties, we are unable to make decisions having a sufficient quality. Also if the range of data that are available is too restricted, they cannot be used as the basis of a decision intended to satisfy certain requirements.

In decision models it is very important that it be possible to determine *the quality of any particular decision procedure*, i.e. to express the measure of its adequacy within the framework of a certain problem, or class of problems, of decision-making. We are led in this connection to refer to the concept of 'risk' in the sense as used in the theory of statistical decision.[137] Here the concept of 'risk' is understood in the broadest sense, that is so as to include what might be described as the degree of approximation attainable with the use of a given decision procedure, such as a mode of explanation, prediction, method of diagnosis, and suchlike.

It is clear from these considerations that decision models of scientific procedures operate, on the one hand, with elements that we are able to reconstruct from inferential models of scientific procedures (initial data, resultant data, i.e. input and output space, the communication channel including the relevant transformational procedures based on knowledge of the general relations of the elements of the output space to elements of the input space or vice versa), and on the other hand certain additional elements which make it possible to assess the quality of decision-making and the relevance of the means operated with.

Abstract models of decision-making have been elaborated in some considerable detail in the theory of statistical decision functions[138] or that of strategic games. The search for the optimal performance or solution of scientific procedures can be treated as a problem of statistical decision, or as a game between two players, where one is nature or any characterized universe, and the other man (a scientist, doctor, or anyone endeavouring to carry out the given scientific procedure). If we imagine the decision process as a game with nature, we cannot treat nature as an opponent whose actions are deliberate and who deliberately tries to take advantage of the errors committed by man in the course of the process.[139] In the following account, therefore, we shall not be looking for any new conception of a decision model, but merely indicating how such models are to be interpreted in accordance with the language in which we have been describing the scientific procedures of the type in question. Abstract models of decision operate with the following elements: [140]

Data of an empirical nature, e.g. data on the state of a patient and the symptoms that have been ascertained, data on causal sequence or antecedence (depending on whether we are to determine antecedence or sequence), data on the present state of the system whose future state we are to determine, etc. Such data we shall regard as the *output, or selection, space* Y. This selection space can also be treated as the set of all the results of the experiments that the scientist must make, or as the set of all the possible items of information that can be received at the output of the information channel.

Further there is what we might call the *input space* X, which corresponds to the resultant data of a scientific procedure, e.g. the states of a system that we are to determine, the set of all possible illnesses or pathological states in diagnostic procedures, data on causal antecedence or sequence (depending on which we are to ascertain and which of them is already taken up as an element of space Y). Input space can also be regarded as 'parametric' space, or as the set of all possible states of nature, or of the universe, which are to be determined. Regarding the abstract decision model as a model of a communication process, then we take the input space to correspond to the set of all possible states in the source of information. If we take as our basis the conception of a scientific procedure as a game between man (the scientist, doctor, etc.) and nature, every $x(x \in X)$ represents a strategy of nature.

A further element in the abstract decision model is the *decision space* D which in effect represents the set of alternatives from which man may select on the basis of the data of the output space. In the communicational interpretation, this is the classical problem of detecting what piece of information has been emitted, on the basis of the information received in the output space. From the point of view of scientific procedures of the kind that interest us, D can be regarded as an appropriately selected subset of X, which means, for example, the states that we are to determine, the set of illnesses or possible illnesses and remedial procedures that the doctor has at his disposal at a given moment, etc.

The most significant element of the abstract decision model is the space of the *decision functions* F. In effect these are the set of rules that enable us to select an adequate decision on the basis of the general interrelations ascertained between the input and output spaces. In the interpretation of the decision process as a game between man and nature, F represents the set of man's pure strategies against nature. Each element of space F represents a procedure which enables us to assign a certain $d(d \in D)$, or a certain distribution of probability in D to any $y(y \in Y)$. In the first case we speak of a pure or *non-randomized* decision function, in the second we speak of a mixed or randomized decision function. In the case of prediction, each f $(f \in F)$ represents a procedure whereby we are enabled to determine future states (which we regard as elements of D, or X as the case may be) on the basis of states that have already been ascertained (and which we regard as elements of Y). In diagnostic procedures, every f represents one possible diagnostic approach with respect to the given population, the given set of illnesses, or available remedies, as the case may be, as well as with respect to the set of means that we have at our disposal for ascertaining the symptoms.

As we have already pointed out, decision functions are a way of processing our knowledge of the general interrelations between the elements of the input and output spaces. This also means that in the decision model we consider scientific laws, scientific hypotheses – irrespective of whether they are of the deterministic or statistical type – as well as all the corresponding knowledge (e.g. medical knowledge in diagnosis) as appropriate decision functions.

In the light of these elements of the abstract decision model it is also possible to give a different interpretation of the concept of 'explaining'. In inferential models this concept is understood roughly in the sense that to explain a given fact means establishing such nomological statements, as well as statements on certain concrete conditions, that it is possible to infer a statement about the given fact from the nomological statements and those concerning concrete conditions. In decision models of scientific explanation the concept of 'explaining' is understood as follows: to explain a certain y means finding such a decision function f $(f \in F)$ as makes it possible to assign to each y a certain d or a certain distribution of probability in D. The above-mentioned requirement that D be a subset of X should be understood as meaning that the explanation is to be 'real', i.e. it is to operate with objects, properties and relations that can be shown to exist in nature. (It is of course possible to imagine explanations that do not operate with objects, properties and relations answering these requirements, e.g. explanations operating with such concepts as 'phlogiston', 'supernatural beings' and the like. In that case the requirement that D be a subset of X is not fulfilled. Thus we must always allow for such apparent explanations where the intersection of D and X is empty. We might speak in such cases of 'seeming explanation', 'illusive explanation', and so forth.)

If we were to take into account only those elements of the abstract decision model that we have considered so far, we should not be able to evaluate the quality of the decision-making. We should not, for instance, be able to distinguish between a relatively more and relatively less accurate prediction, a better or worse explanation, a more or less adequate diagnosis. This calls for certain other elements to be considered in the abstract decision model, elements which would furnish criteria for evaluating the quality of our decision-making.

The first thing we must respect is that the input space X is not accessible immediately but only via the data which we obtain at the output of the *communication channel* (the latter may also be described as the 'channel of science', the 'observer channel', 'experimenter channel', etc.). In other words 'nature's strategy' is not always of ready access to man, who must take steps to discover its actual character gradually. This channel, interpolated between the input and output spaces, causes what is described in the theory of information as noise, which means the possibility of error arising in the transfer of data from the input to the output space owing, for example, to certain properties of the channel, insufficient fineness in differentiating, or the insufficient standard of the measuring and experimental apparatus, for instance apparatus for distinguishing individual symptoms etc. It would be ideal if we could break the space Y down into a collection of disjunct sets such that each of them would correspond to a certain x. In that case it would become possible to speak of perfect discriminability during the transfer through the channel. In fact, however, all the real channels with which we operate only approximate to perfect discriminability to a greater or lesser degree. In general we assume that a certain distribution of probability is characteristic for the space X, and that the relations between the elements of X and Y are also of a probabilistic character.

In order to be able to assess the quality of our decision-making, we must be in a position to evaluate the loss or gain arising out of a decision d if a certain element x was at the input. For the assessment of this loss (or gain), it is expedient to introduce what is called the *weight or loss function* $w(x, d)$, which assigns to each pair $\{x, d\}$ a certain loss, where x is a state which really did appear at the input, and d is the selected decision. This loss might be understood, for example, as the deviation of the actual properties of a predicted state from the properties which emerged as the result of the prediction, or as the aggregate of losses (or gains) which arises with the selection of decision d, if the real state x was at the input. The concept of weight or loss function will be seen to be defined on the Cartesian product of X and D. Intuitively it is also clear that the less adequate the selected decision d with respect to the real state x, the greater must be the corresponding loss $w(x, d)$.

It still remains to introduce the *risk function*, which is defined on the Cartesian product of X and F, i.e. as the function which assigns to each pair $\{x, f\}$ the total loss incurred when with state x decision function f has been used on the basis of y. Taking into account all the elements of the abstract decision model, we describe the problem of decision-making as the task of finding the optimal decision function such as the risk is minimal. Procedures have been worked out which look for the optimal decision function in the minimization of the average risk (the Bayes solution), or there are procedures which minimize the maximal risk (the 'minimax solution').[141]

As will be apparent the abstract decision model operates with certain pragmatic aspects which make it possible to assess the quality of decisions and thereby the power of the respective decision functions. There is also the possibility of considering the relevance of the different elements with which we operate, as well as the reduction of those elements which either are irrelevant or whose omission has little or no effect on the quality of our decision-making.[142] On the other hand, if the quality of our decision-making is unsatisfactory, it becomes necessary to extend the collection of elements, e.g. extend the output space, to extend and improve the space of decision, the space of decision functions, etc. Moreover it should again be recalled that, like the reduction procedures, procedures leading to the extension of the collection of elements must also be regarded as relative, i.e. as bound to the demands placed on the quality of the decision-making, and with respect to the admissible risk level, and so on.

If we make the quality of our decision-making, which means the quality of an activity performed and controlled by man, a criterion for the admissibility of this or that reduction of the elements with which we operate in a given procedure, or a criterion for assessing the necessity of their extension as the case may be, the objection might be raised that this amounts to a kind of pragmatization of the categories with which we operate in scientific procedures. It might also seem that we are returning in this way to principles similar to those which lay behind the various operationalist or instrumentalist conceptions of scientific laws. There is, however, one major difference: while the operationalist or instrumentalist conceptions were based on the activity of a human subject, which activity they assessed and analyzed, the conception based on the decision model starts from the idea that there are always *two partners in a game*, namely man and nature. Man as one of the partners does not make his steps independently of those of the other partner. This also means, for example, that the decision model, as well as the other elements which man uses, is not, and cannot be set up by him at liberty, but only in accordance with the level of his knowledge of the strategy of nature. Thus the decision model deliberately calculates with the criterion of practice whenever a judgement is to be made as to the adequacy of the elements which man selects with respect to the strategy with which nature operates.

6. Explanation and Prediction

(a) *The Scientific Basis of Prognostic Statements and the Thesis of the Structural Identity of Explanation and Prediction*

The prognostic statements which we work with in the empirical sciences are usually connected with a combination of epistemic and temporal modalities. The former can be graded according as we assert what we assert as certain or verified, almost certain, highly probable, etc. This can be combined with an explicit time specification, which means that it is asserted (as certain, almost certain, highly probable, etc.) that 'something' will be, that there will be a certain fact or event at a time after the time when the assertion is uttered. If we assume that the fact takes place at time j while the assertion is uttered at time i, then we can distinguish

Prediction, if i precedes j in time,
Retrodiction, if j precedes i in time,
Codiction, if i and j are simultaneous.

This distinction between prediction, retrodiction and codiction is of course based on the assumption that the respective time differentiation of i and j is at all possible. Allowance must always be made for the fact that the expression 'will be' is ambiguous in a way similar to the expression 'is'.[143] It may relate to a limited period, or to the unlimited period commencing with the moment i, or it may also be of a timeless character. In the present analysis of prognostic statements we shall confine ourselves to the situation where j relates to a limited and strictly defined period, so that there is full justification in considering the time relation between i and j.

Prognostic statements may be of the most various nature. Man has always had a deep interest in the future, and has sought both rational and irrational means for its prediction. The various oracles of the past, the augurs and haruspices of ancient Rome, as well as certain writers in our own day have all had ambitions to produce statements about future events, but prognostic statements of this kind – we call them prophesies – have scarcely anything to do with science. In our account we shall confine ourselves to rationally justified predictions of future events, i.e. those that can be reconstructed on the basis of rational argumentation. [It goes without saying that there is no sharp dividing line between prophesies[144] and scientifically justified predictions, and that the criteria by which we distinguish between them are historically conditional and variable.]

All the authors who have pointed out the deductive character of schemata of scientific explanation have also drawn attention to the connection between scientific explanation and scientific predictions. On the basis of the schemata it is possible to deduce a prognostic statement from known scientific laws and the relevant empirical conditions in the same way as it is possible to deduce an explanandum. This conception lies behind the well-known thesis of the structural similarity of explanatory and predictive procedures. Hempel and Oppenheim formulated the thesis of the structural similarity of explanatory and predictive procedures as follows: The same formal analysis, including the four conditions of adequacy, apply to both explanation and prediction.[145] The difference between the two procedures is purely pragmatic. Hempel later presented a new form of the thesis, namely as the conjunction of two sub-theses: (a) Any adequate explanation has a potential predictive character, (b) any adequate prediction is potentially an explanation.[146] The introduction of the concept of 'systematization' as *genus proximum* for all such procedures[147] is also based on the structural similarity of explanatory and predictive procedures (as well as the procedures of retrodiction and codiction). From this point of view systematization is a rational combination of statements some of which have the character of nomological statements enabling us to decide as to those statements which come under the semantic competence of the nomological statements used. This means that systematization is also represented by diagnostic procedures, some procedures of classification, procedures of derived measurement based on the use of nomological statements, and others.[148]

Although the concept of 'systematization' is a doubtless useful generalization for a certain class of scientific procedures, the differences between the procedures are easily shown to be more profound and irreducible merely to pragmatic aspects. Accordingly objections can be raised to Hempel's thesis of the structural similarity of explanatory and predictive procedures, or the two sub-theses as the case may be.

The assertion that any adequate explanation has a potentially predictive character is either a simplification or an unremunerative narrowing-down of the concept of 'adequate explanation' to the explanation of relatively isolated dynamic systems whose behaviour can be explained by deterministic laws.[149] It is possible to explain, for example, why a given person or group of persons became adherents of a certain opinion. Yet this explanation will usually not enable us to determine with accuracy what views the given person(s) will hold in the future, the less so if we are not acquainted with all the further circumstances that might influence the development of an opinion. We always demand of an adequate scientific explanation that it be a *relatively dependable basis for decision-making*, it being taken for granted that that to which the decision-making relates may include both present and future events. Also, the measure of dependability of this basis may vary, not only from case to case of explanation, but also in the same explanation with respect to present or future events.

Some authors have also expounded the thesis of the structural similarity of explanatory and predictive procedures on the basis of their symmetry.

Taking the basic schema $C.T \rightarrow E$, then the procedure from a given C and T is a prediction (or also retrodiction or codiction), and the procedure from E to $C.T$ is an explanation. The symmetry is of course merely apparent. It would be possible to speak of a certain relative symmetry if in addition to C, or E, we were also given T. In that case, of course, the explanation is the finding of causal antecedence (and coincides with retrodiction), and prediction is the finding of a causal sequence. Hence it is clear that the thesis of the symmetry of predictive and explanatory procedures fully conforms to the principles of Laplace's determinism whereby the laws of motion are given and known *a priori*. Yet if we consider that the result of a scientific explanation is not only the discovery of C but also that of T, then the assumption of symmetry breaks down. One part of the result in procedures of explanation is made up of nomological statements in addition to singular statements, and in procedures of prediction the desirable result takes the form of singular statements about future states.

The significant difference between the results of procedures of explanation and prediction will also become clear from a comparison of the semantic characteristics of what corresponds in the latter to the explanandum in the schema of the former. Explanation assumes the explanandum to be the assertion of a true statement, the latter being related to an established, demonstrable and verified fact in the past or present. A statement about a future fact or event cannot be true in the same sense as a statement about a present or past fact. It is at least scarcely possible to imagine procedures of verification which would correspond to those we are accustomed to and would still be capable of deciding prognostic statements.

In the semantic analysis of prognostic statements it would seem to be necessary to reconstruct a universe of possible facts, or a universe of alternative possible worlds. If prognostic statements are based on a certain scientific theory, it is profitable to consider whether this or that fact, or this or that alternative possible world, is possible with respect to the scientific theory accepted. So the semantic analysis of prognostic statements involves decision-making *de futuris contingentibus*, where the decision-making must be relativized to an aggregate of assumptions, e.g. to an accepted and confirmed scientific theory about the domain in which the prognostic statements are made. Hence there is naturally a semantic difference in the nature of statements of the kind that the sun will rise tomorrow at such and such o'clock, that the next complete eclipse of the Sun will be at such and such a time and will be visible at precisely such and such a place, etc., statements based on statistical laws, such as those of half-life, etc., and finally statements based on less rational assumptions, such as those relating to certain events of private life. From both the practical and theoretical points of view, it is expedient to distinguish between singular prognostic statements which we relate to future possible facts or events, and general prognostic statements. Even general prognostic statements may be of great value in practice and in theory, as in the case, for example, of statements to the effect that all the minerals which may be found on the Moon will have such and such properties, etc.

So the semantic analysis of prognostic statements cannot be identified with the ordinary analysis of indicative statements relating to present or past facts. Hence we cannot speak in terms of the truth of prognostic statements in the strict sense of the word. On the other hand, however, it is mandatory to respect the fact that even prognostic statements can be qualified to some purpose from the semantic point of view.

In an effort to avoid these objections against the thesis of the structural similarity of explanation and prediction, objections based on the different semantic characteristics of the explanandum on the one hand and prognostic statements on the other, Hempel introduced the concepts of 'potential explanation' and 'potential explanans'. This modification made it possible to abandon the fourth condition of adequacy, according to which the statements that make up the explanans must be true. The modification fails in its purpose, however. The point in question is not the moderation of the demands on the explanans but that account must be taken of the fact that there is a fundamental semantic difference between an explanandum and a prognostic statement. Hereby Hempel's concept of 'potential explanation' fails to reinstate the symmetry of procedures of explanation and prediction.

However, the asymmetry of explanation and prediction is even broader and does not apply merely to semantic circumstances. It also relates to circumstances of a pragmatic nature: there are generally two things which we expect of an adequate scientific explanation, namely the explanation of a given event, and the explanation of a certain class of possible events similar to that which has been ascertained. The measure in which a scientific explanation meets the second expectation then determines how far the explanation may serve as a reliable basis for decision-making, i.e. for procedures of prediction. Thus we may conclude that Hempel's thesis that any adequate explanation has a potentially predictive character is not acceptable, not only for semantic but also for pragmatic reasons. This is not of course meant as saying that we deny any connection between explanation and prediction, but merely that no connection can be set up universally for all cases of explanation.

Similar objections might be raised to the second of the sub-theses, which asserts that any adequate prediction is potentially an explanation. It is usually pragmatic yardsticks that are authoritative for assessing the quality of individual predictions. Moreover, even in everyday conversation we speak of more or less successful predictions. The common measure of successfulness of prognostic statements is in the agreements or differences in the parameters of the predicted events and the events that actually occur. However, the successfulness of a prediction is no guarantee that the predicter has available a sufficiently reliable basis for prediction or that he is fully acquainted with some such basis.

This possible difference between the possibility of successful prediction on the one hand, and successful scientific explanation on the other, is only a part of a broader set of circumstances involving rational decision-making. Decision-making of this kind may be successful not only because the person making the decision has a complete rational command of all the processes on which the given decision process is based, but also because of the availability of dependable mediated data. The rationality of our decision-making does not always require immediate knowledge of the objective processes to which the decision-making relates, but may also be based on mediated data from reliable specialized sources, much as the successful driving of a motor vehicle does not require a detailed knowledge of the cycles of internal combustion engines. Similarly a highly successful prediction of a future event may be made on the grounds that the competent specialists agree on like prognostic statements, e.g. the accurate prediction of certain astronomic phenomena may be made because a certain number of qualified astronomers all make the same prediction.

It is true, of course, that it would be necessary, as Stegmüller points out,[150] to distinguish two forms of why-question relative to future events:[151] (1) 'Why will event *x* take place at time *r*?' and (2) 'Why are you convinced that event *x* will take place at time *t*?' The former must be understood as demanding the explicit specification of the basis for decision, or prediction as the case may be, while the latter makes no such demand, asking merely for the justification of the conviction, belief, etc.[152] This differentiation of two forms of why-questions related to future events is no more than the differentiation of abstract types. In concrete procedures of prediction is it not only possible, but even advantageous to link or even combine the two types. This is after all confirmed by practice in social and technical prognosis, which is based on the immediate determination of causal series, procedures operating with certain laws or at least statistical hypotheses, and on the judgments of competent specialists, without the explicit specification of the respective basis for prediction. [Of this kind are, for example, statements based on what has been called the Delphi method, or other methods exploiting the judgment of the competent specialists.[153]] So modern prognostic practice simply does not confirm Hempel's thesis that any adequate prediction is potentially an explanation. This would mean that we can make successful predictions of future events even without being able to explain these or like events scientifically.

Prognostic statements may be successful not only because use is made of the judgment of competent specialists who, moreover, are in a position to furnish a scientifically acceptable basis for the predictions and the appropriate scientific explanations, but also, sometimes, even if no such explanation is known. For prognostic statements may also be based on certain empirical generalizations, e.g. the extrapolation of empirical trends hitherto discovered, without there being any known scientific explanation for those trends. In such cases a certain basis for prediction is afforded by the assumption of the stability of certain characteristics, tendencies or a given sequence of changes. This means *de facto* that the only basis here is empirical knowledge of the present and its projection into the future. Such procedures are naturally only very limited and also have no more than a limited reliability.

The thesis of the structural similarity of explanation and prediction is thus unacceptable even in the form of the second sub-thesis. Generally speaking there is a relatively wide circle of explanations that are not potentially predictive, e.g. the explanation of death owing to the advanced state of a certain illness, the explanation of the break-down of some technical apparatus owing to wear having reached a certain level, the explanation of the collapse of a building owing to an earthquake of a certain intensity, etc. These and similar explanations would be potentially predictive if it were true that every generally defined state of a given illness leads to death, that any wear in the components at a certain defined level would inevitably lead to the breakdown of the apparatus, etc. These and similar explanations clearly cannot usually be interpreted in this way. We might also point to the wide range of prognostic statements which do meet the required demands of accuracy or successfulness without being based on any immediately available scientific explanation, e.g. those based merely on the judgments of qualified specialists, empirical generalizations, the extrapolation of known empirical trends, etc.

This criticism of the thesis of the structural similarity of explanation and prediction must not be understood to mean that we deny that there is any use in scientific explanations with respect to the quality and successfulness of prognostic statements. The measure of their usefulness may vary, however. We have said earlier that there are two things expected of an adequate scientific explanation: (a) the explanation of the given event, (b) the explanation of the whole class of possible events similar to that in question. If a scientific explanation can only satisfy (a) as in the case of the foregoing examples of explaining a death, or a breakdown in a piece of machinery, where it is not usually possible to establish the generally valid level at which the phenomena given as causes actually manifest themselves, no completely reliable conclusions can be drawn from these explanations for making prognostic statements.[154] This is the case with numerous scientific explanations. Darwin's theory of natural selection, for instance, gave a satisfying explanation of the emergence of species that already existed, but it is not a sufficiently reliable basis for prognostic statements which would predict the origins of new species. In so far as an explanation also meets expectation (b) above to a sufficient degree, then far from being merely possible, it is indeed mandatory to take this explanation into account in the formulation of scientifically justified prognostic statements.

(b) *Prognostic Statements and Problems of the Time Factor*

As we have shown, the quality or successfulness of prognostic statements cannot be reduced to the question of whether we have the appropriate basis for prediction immediately at our disposal. Their quality also depends on the objective nature of the systems to the future states of which the prognostic statements relate. Up to the second half of the 19th century one pattern of scientific thinking was the kind of analysis of an isolated deterministic system the elaboration of which is associated especially with the names of Copernicus, Kepler, Galileo and Newton. This kind of analysis reached certain peaks in the astronomy of the solar system. Some maximal demands on the possibilities afforded by the analysis of such a system are included in Laplace's familiar formulation of classical determinism. The absolute intelligence which this formulation assumes can then be envisaged as the model of the perfect predicter.

The evolution of physics and the other branches of science since the second half of the 19th century, in particular the development of statistical physics, thermodynamics, and later the results of the quantum theory and certain other branches, has shown that this pattern of scientific thinking, the model of which was Newton's astronomy and Laplace's formulation of classical determinism, not only has its limitations but is also inadequate in some domains. These are domains where it is not possible on the basis of known scientific laws to predict the future states of a system, but only the probability of possible future states, the probability having moreover a falling tendency in proportion to the remoteness from the present of the future that is to be predicted. Wiener considers two types of scientific thinking in this connection. He describes the first as the type of Newtonian astronomy, the second as the type of the meteorological system.[155] The former allows the making of prognostic statements marked by a high degree of accuracy, the actual degree of accuracy being independent of the remoteness of the time of the event predicted from the time-base, i.e. the moment when the prediction is made, or the time from which are taken the initial empirical data for calculation. In contrast, the second type allows of the production of probabilistic statements, the probability falling with an increase in the distance in time from the time-base. Wiener also points out that scientific thought tends to develop in the majority of the natural sciences somewhere between the two types, though the majority of all sciences come closer to meteorology than Newtonian astronomy.

These thoughts rather confuse a number of different criteria which ought in fact to be kept strictly apart in the observation of the quality of prognostic statements. We have in mind the following criteria, which are generally conceived in their extreme, polar forms. (Their actual form may of course fall somewhere between the poles.)

(a) The quality of prognostic statement depends on how far the system in question is a *relatively isolated system* or one which is open to interventions from without, or one where allowance must be made for such possible interventions including those which cannot be foreseen.

(b) The quality of a prognostic statement depends on how far the behaviour of the system in question can be explained by *dynamic laws* or *laws of a statistical nature*.

(c) The quality of a prognostic statement further depends on how far it is possible in the given system to assume *homogeneity and isotropy* of space or how far this homogeneity and isotropy is limited.

(d) Another criterion which must be taken into account is how far it is possible in the given system to assume *homogeneity of time*, or how far this is limited.

The concept of a 'relatively isolated system' is of course a mere abstraction to which real systems may be more or less close approximations. Wherever it is possible to ignore, in practical calculations, negligible exterior influences and rely in explanation and prediction on dynamic laws, as in the case with Newtonian astronomy (where, moreover, it is also possible to assume homogeneity of time and space), it becomes possible to make highly accurate predictions. It is usually held in these cases that the accuracy of the predictions does not depend on the actual properties of the systems being analyzed but on the accuracy with which the initial conditions have been specified, i.e. the accuracy of observation and measurement etc. In systems of this kind we also encounter another significant circumstance which we might describe as the symmetry of retrodiction and prediction. Since the nature of the time and space which we can take into account is practically unvarying, and since we are in effect operating with scientific laws where the absolute form of time and space parameters is irrelevant, the procedures of prediction and retrodiction are practically identical. Moreover, these procedures practically coincide with derived measuring, with the sole difference that the latter usually operates with laws of coexistence, such as Boyle's law, the laws of geometry whereby it is possible to calculate the sides of a triangle on the basis of known angles, etc. In contrast prediction and retrodiction operate with laws of sequence, i.e. laws which calculate with different time and space relations, only relative time and space differences being taken into account and absolute time and space parameters being regarded as irrelevant.

The symmetry of prediction and retrodiction is of course only a borderline case if the above-mentioned conditions are fulfilled. It is sufficient to take those physical systems where we must allow for irreversible actions, such as those of thermodynamics, and for entropic actions, and we see that there can be no assuming symmetry of prediction and retrodiction. And it is even less possible in open systems such as those of living nature, social or socio-economic systems, and suchlike. In these cases time and space are not homogeneous, and the systems undergo different developments in different time and space conditions.

The difference between the symmetry and asymmetry of prediction and retrodiction, and simultaneously the different role of the time factor in the past, can be illustrated by means of Figures 8 and 9.

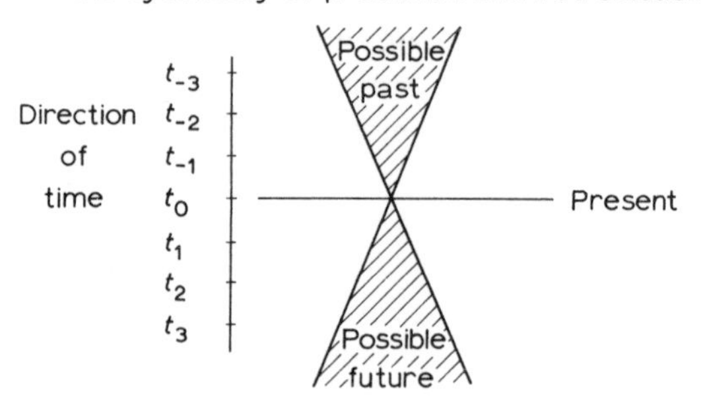

Fig. 8.

Figure 8 shows that quite definite states in past or future correspond to a certain state in the present – in the case that deterministic laws can be applied – or that certain distributions of the probability of states in the future or past correspond to the present state – in the case that statistical laws can be applied. The shaded part indicates possible past and future, the choice of the initial moment t_0 being entirely a matter of convention and in no way influenced by the nature of the system. (In the case of the usability of deterministic laws the shaded area coincides with a straight line, since to every initial state at time t_0 there is a quite definite corresponding state at times t_1, t_2, etc. in both the future and the past.

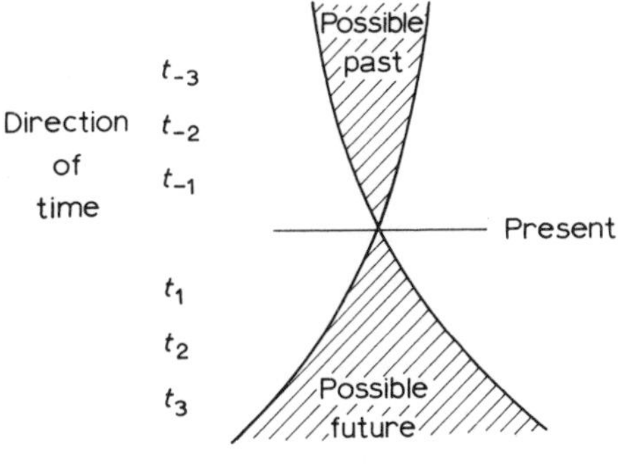

Fig. 9.

Figure 9 also shows that to each state in the present there is a certain corresponding distribution of probability of states in the past, but a different distribution for states in the future. This means that the aggregate of the changes that take place during a certain physically characterized time interval is not the same in the different periods of time. In systems where there is a growth in complexity and the measure of their organization, the role of the time factor is in accordance with this diagram. A typical example of such a system is afforded by the systems of biology, or social and socio-economic systems. In these diagrams, time may also be regarded as the external background to all action, as in Newton's conception of time. In fact, however, what takes place in historically different but physically the same time intervals is not the same and comparable, so that we cannot speak in terms of the homogeneity of time or the same role of the time factor in all historical situations.

There is also no difficulty in imagining systems whose complexity and measure of organization decreases, i.e. systems with increasing entropy. Also in such systems it is not possible to consider homogeneity of time and the same role of the time factor, although the diagrammatic expression of the relations between 'possible past', 'present' and 'possible future' could in that case be mapped in the reverse.

The gauge of the non-homogeneity of time (which must not be regarded as an abstract stable background but as an inherent form of the motion of a real system) is the half-life, i.e. interval, in which the scope of the system, which is being somehow measured, is doubled or halved. This method, which is successfully applied in physics for characterizing certain changes or energic processes, can also be used in the case of systems of a non-physical nature, although in such cases the given interval is usually variable.[156]

Whenever a system under investigation is an open one and must be considered historically, i.e. non-homogeneity of time must be respected, this also means that prognostic statements which meet the demands of quality and successfulness are possible only up to certain time-limits, their quality decreasing the closer the limits are approached.

We have already indicated that a distinction should be drawn between systems whose evolution cannot be affected by intervention (irrespective of whether they can be characterized by dynamic laws as in the case of Newtonian astronomy, or statistical laws as in the case of meteorology), and systems where intervention is to a greater or lesser degree possible. If it is possible to determine the nature and effects of the interventions, especially those organized and controlled by man, there is the possibility of formulating what we might generally describe as *conditional prognostic statements*. The difference between simple and conditional prognostic statements will be clear from the following schematic examples:

(1) With the mathematical expectation $r(0 \leqslant r \leqslant 1)$ we expect the system S at time t, which follows the time when this statement is made after an interval of Δt, to be in state S_t.

(2) With the mathematical expectation $r(0 \leqslant r \leqslant 1)$ we expect that if at time t, which follows the time when this statement is made after an interval of Δt, the events E take place (or if we succeed in causing them, if this was our aim), then the system S will be in the state S_t.

Statement (1) is a simple prognostic statement, statement (2) a conditional prognostic statement. It will be apparent that a conditional statement corresponds to the formulation of counterfactual conditions (especially those in the form of unreal conditionals), but with the difference that the epistemic and temporal modalities related to future events are stated explicitly. This means then that here – as in the case of the ordinary formulations of counterfactual conditions – we operate with certain *a priori* knowledge of the real interrelations of possible events, though neither the antecedence taken in isolation, nor the sequence taken in isolation can be considered true.

For the formulation of conditional prognostic statements the relevant *a priori* knowledge, especially knowledge of the adequate scientific laws, is of paramount importance. Since conditional prognostic statements are a major component of any rational planning or decision-making orientated towards a possible future generally, such planning or decision-making is unthinkable without knowledge of the adequate scientific laws.

(c) *Justified Prognostic Statements and Their Evaluation*

In addition to the distinction between simple and conditional prognostic statements, a further distinction should be drawn between unjustified and justified prognostic statements. The difference follows from the following examples:

(1) We expect that ....
(2) We expect that ..., because ---.

The expression ... is a prognostic statement, irrespective of whether simple or conditional, and the expression --- is its justification.

The conception of justified prognostic statements may serve as the basis of the semantic evaluation of prognostic statements. If we record a justified prognostic statement in a way corresponding to Hempel's models of systematization, we obtain

$$S_i \text{ because } S_h . S_e,$$

where S_i is a prognostic statement, S_h a nomological statement, and S_e the statement of certain empirical initial conditions. The semantic evaluation of the prognostic statement S_i, which relates to a future possible fact, can be made with respect to S_h and S_e, which are available *hic et nunc*. This also means that this evaluation of prognostic statements is relativized to the epistemic means with which we are in a position to operate at the present time. It further means that the semantic evaluation of prognostic statements – as thus relativized to the present – may be modified if in the future there is a change in what we may regard as the predictive basis [157] for S_i.

As far as concerns the evaluation of the predictive basis, one possibility is the application of the concept of 'systematic (explanatory, predictive) power'.

The definition of the measure of systematic power of a nomological statement S_h with respect to S_i can be modified by counting, in addition, with the initial conditions S_e as follows:

$$E(S_h, S_i/S_e) =_{df} \frac{U(S_i/S_e) - U(S_i/S_h.S_e)}{U(S_i/S_e)}.$$

Intuitively it is clear that in the justification of a prognostic statement there ought only to appear the nomological statements and statements of the initial conditions that are relevant with respect to S_i. If, for the sake of simplicity, we leave S_e to one side (by, for example, including it in S_h), we may describe the problem of the relevance of the individual components which figure in the predictive basis more accurately thus: Since the measure of systematic power is additive for any nomological statements that are logically disjunct, it holds, provided S_h and S_k meet this condition, that

$$E(S_h.S_k, S_i) = E(S_h, S_i) + E(S_k, S_i).$$

In connection with this condition for the additivity of the measure of systematic power it should be noted that there are certain consequences that are of importance for establishing the *optimal prediction basis*. It is clear from the above condition that for n nomological statements which are mutually logically disjunct it holds that

$$E(S_{h_1}.S_{h_2}....S_{h_n}, S_i) = E(S_{h_1}, S_i) + E(S_{h_2}, S_i) + \cdots + E(S_{h_n}, S_i).$$

If $S_{\text{theor.}}$ is the conjunction of the nomological statements characterized in this way, it is expedient to look for all the components of this conjunctive class of nomological statements that are capable of increasing the systematic power with respect to what S_i relates to. At the same time it is possible to formulate the *rule for reduction*:

From $S_{\text{theor.}}$ can be excluded all nomological statements that do not increase the systematic power of $S_{\text{theor.}}$ with respect to S_i. In other words we can exclude all the nomological sentences of which it holds that

$$E(S_{h_j}, S_i) = 0 .$$

This rule for reduction can also be formulated as follows: all the nomological statements that are exclusively disjunct with S_i, i.e. those for which it holds that $E(S_{h_j}, S_i)=0$, may be described as systematically irrelevant with respect to S_i. Then the rule for reduction has the form: From $S_{\text{theor.}}$ as the predictive basis for S_i can be excluded all nomological statements that are systematically irrelevant with respect to S_i.

In establishing the optimal predictive basis for S_i it is worth formulating a requirement which is a kind of analogy to Carnap's familiar requirement of total evidence for probabilistic and inductive logics (Carnap, 1950; p. 211). If in establishing the degree of confirmation of a given nomological statement it is expedient to pronounce the requirement of total evidence, or some less strict variant of it (e.g. the requirement of maximal actually, i.e. *hic et nunc*, available evidence), it is equally expedient in establishing the optimal prediction basis to formulate the following requirement:

(RI) *In determining a suitable prediction basis with respect to S_i all knowledge that is not systematically irrelevant to S_i must be taken as its foundation.*

On the strength of the foregoing considerations on the predictive basis it is possible to introduce certain other concepts, primarily that of the 'permissible reduction' for $S_{\text{theor.}}$ and that of 'sufficient predictive basis'. Let us imagine that $S_{\text{theor.}}$ is a conjunctive class of nomological statements that are mutually logically disjunct. Then by the permissible reduction for $S_{\text{theor.}}$ with respect to S_i we understand the omission of those elements of $S_{\text{theor.}}$ which are only negligibly systematically relevant to S_i, those where

$$E(S_{h_j}, S_i) \leqslant \varepsilon ,$$

where ε is a conventionally agreed value not very different from 0.

The concept of a 'sufficient predictive basis' with respect to S_i can be introduced in a similar manner: $S_{\text{theor.}}$ is a sufficient predictive basis with respect to S_i if it holds that

$$E(S_{\text{theor.}}, S_i) \geqslant \pi ,$$

where π is a conventionally agreed value which ensures a sufficiently reliable level of what S_i relates to.

The requirement (RI) can be modified according as it is based on the modified measure of systematic power. In that case it becomes necessary, in determining the predictive basis with respect to S_i, to take into account not only $S_{\text{theor.}}$, which includes all the nomological statements that are systematically relevant with respect to S_i, but also all the statements about the initial conditions that have been empirically established, that are systematically relevant with respect to S_i. Then the term 'all knowledge that is not systematically irrelevant to S_i' includes not only $S_{\text{theor.}}$ but also the total available evidence. This requirement so understood, as a requirement for determining the optimal predictive basis, is in effect an inductive logical analogy of Laplace's famous conception of an *intelligence absolue*.

Our analysis of the semantic informational potential of justified prognostic statements has so far been related mainly to what we have described as the predictive basis with respect to a prognostic statement S_i. However, as the point of departure for evaluating S_i we might well select a different procedure whose intuitive basis is the following consideration: Since the role of S_h, or $S_h.S_e$ as the case may be, as predictive basis with respect to S_i is the reduction of uncertainty with respect to that to which S_i relates, we might ask how far S_i is supported by the fact that we have S_h, or $S_h.S_e$, at our disposal. The fact of how far S_i is supported by our having S_h, or $S_h.S_e$ respectively, at our disposal might be described as the *nomological support* of S_i, or its *nomologico-empirical support*.

If $N(S_i/S_h)$ is the nomological support of S_i on the basis of S_h, we might consider the measure of nomological support $N(S_i/S_h)$, or the measure of nomologico-empirical support $NE(S_i/S_h.S_e)$ as the case may be, which would fulfil the following requirements:

(r1) if $S_i \underset{L}{\equiv} S_j$ and $S_h \underset{L}{\equiv} S_k$, then $N(S_i/S_h) = N(S_j/S_k)$,

(r2) if $S_h \underset{L}{\rightarrow} S_i$, then $N(S_i/S_h) = \max N$.

The latter requirement allows of the addition that if it is desired that (in accordance with the mode of quantification used so far) the value of $N(S_i/S_h)$ vary between 0 and 1, then max $N=1$. In accordance with this the next requirement is selected:

(r3) if $S_h \underset{L}{\rightarrow} \sim S_i$, then $N(S_i/S_h) = \min N$,

(r4) $0 \leqslant N(S_i/S_h) \leqslant 1$.

[Analogous conditions can now be formulated for $NE(S_i/S_h . S_e)$, which differ from (r1) – (r4) in that everywhere in place of the nomological statement S_h there is the conjunction $S_h . S_e$.] These conditions are clearly seen to be met by a concept which is the complement of cont (S_i/S_h), or cont $(S_i/S_h . S_e)$ as the case may be. In other words the concept of nomological support, or nomologico-empirical support respectively, might also be described as the conditional certainty of S_i with respect to S_h, or $S_h . S_e$ respectively. This also means that the concept of 'nomological support', or 'nomologico-empirical support' respectively, may also be used to show the likelihood of the prognostic statement S_i with respect to S_h, or $S_h . S_e$, and the more adequately the more fully the requirement (RI) is met.

The determination of a prognostic statement on the basis of an available prediction basis might also be regarded as a process of decision-making. If we were able in such a process to make decisions with absolute certainty as to the reliability of all the data, there would be no need to bring the risk element into consideration. In fact, however, there are numerous procedures, whose aim is the determination of sufficiently reliable prognostic statements, where it is expedient to make allowance for a certain risk. These reasons lead to the possibility of also considering the applicability of some of the categories of the theory of decision.

The main thing is what we gain in having opted for prognostic statement S_i and not for its negation, i.e. $\sim S_i$. The *expected gain* will be greater the higher the measure of nomological support of S_i on the basis of S_h, or $S_h . S_e$ as the case may be; and it will be greater the higher the measure of systematic power of S_h, or $S_h . S_e$ respectively, with respect to S_i. At the same time we can say that the expected gain will be less the higher the measure of nomological support of $\sim S_i$ on the basis of S_h, or $S_h . S_e$; and it will be less the higher the systematic power of S_h, or $S_h . S_e$, with respect to $\sim S_i$.

If we regard the concept of 'expected gain' as a quantitative concept, then we must respect the possibility of negative gain, i.e. loss. For this reason it is desirable to be able also to calculate with negative values when quantifying this concept. Thus we might consider the following definition of the measure of expected gain (or loss in the case of a negative value), which we denote $G(S_i/S_h)$, or $G(S_i/S_h.S_e)$:

$$G(S_i/S_h) = df\ N(S_i/S_h)\ E(S_h, S_i) - N(\sim S_i/S_h)\ E(S_h, \sim S_i),$$

or

$$G(S_i/S_h.S_e) = df\ N(S_i/S_h.S_e)\ E(S_h, S_e/S_i) - \\ - N(\sim S_i/S_h.S_e)\ E(S_h, S_e/\sim S_i) \text{ respectively}.$$

If we require that this measure of expected gain[158] have values between -1 and $+1$, it is to advantage if we select for the determination of $N(S_i/S_h)$, or $N(S_i/S_h.S_e)$ respectively, the complement of $\text{cont}(S_i/S_h)$, or the complement of $\text{cont}(S_i/S_h.S_e)$, i.e. the likelihood of the prognostic statement S_i with respect to the relevant predictive basis. If likelihood is determined on the basis of the conditional probability $p(i/h)$, or $p(i/h.e)$ respectively, we obtain the following determination for the measure of expected gain:

$$G(S_i/S_h) = p(i/h)\ p(\sim h/\sim i) - p(\sim i/h)\ p(\sim h/i),$$

or

$$= p(i/h.e)\ p(\sim (h.e)/\sim i) - p(\sim i/h.e.)\ p(\sim (h.e.)/i),$$

respectively.

Let us now look at the extreme cases. If it holds that

$$S_h.S_e \underset{L}{\rightarrow} S_i, \ldots$$

i.e. in a dependably deductive nomological case, we obtain

$$N(S_i/S_h.S_e) = 1,$$

$$E(S_h, S_e/S_i) = 1,$$

$$N(\sim S_i/S_h.S) = 0.$$

Irrespective of the fact that in this case it holds that

$$E(S_h, S_e/\sim S_i) \geqslant 0,$$

we arrive at

$$G(S_i/S_h.S_e) = 1.$$

(The expected gain if we opt for S_i, and provided it holds that $S_h.S_e \underset{L}{\rightarrow} S_i$, is then max G, i.e. equals 1.)

However, if we have opted for $\sim S_i$ and it holds that $S_h . S_e \underset{L}{\rightarrow} S_i$, we arrive at

$$G(\sim S/S_h . S_e) = N(\sim S_i/S_h . S_e)\, E(S_h, S_e/\sim S_i) - \\ - N(S_i/S_h . S_e)\; E(S_h, S_e/S_i).$$

Since we arrive at the same values in the reverse order, our loss is maximal, i.e.

$$G(\sim S_i/S_h . S_e) = \min G = -1.$$

If it does not hold that $S_h . S_e \underset{L}{\rightarrow} S_i$, i.e. if the deductive nomological model cannot be used for determining a prognostic statement, as for instance when we operate only with statistical laws, or, to put it another way, if we can draw conclusions about what S_i relates to merely on the strength of probabilistic interrelations, then G may acquire values between -1 and 1.

NOTES

[1] The most significant works are those of Åqvist (1965), Belnap (1963), Harrah (1961, 1963), Kubiński (1966) and Simmons (1967). The last two also contain bibliographies of other literature in the field of erotetic logic.

[2] This viewpoint comes closer to the intuitive use of questions in everyday speech than does the purely syntactic analysis of the system of questions in isolation from statements of an indicative nature.

[3] The author's attention was drawn to these objections and so to the limitations of the approach by Prof. P. Bernays. It must, of course, be seen that any way of recording a problem-solving situation as well as any programmed equipment for answering questions must respect the limitations of the finitistic approach.

[4] More detailed reasons for the differentiation of open and closed questions are given by Giedymin (1964).

[5] In Czech philological writings the terms 'zjišťovací otázky' (questions of ascertainment) and 'doplňující otázky' (complementary questions), respectively, are used. The same distinction is expressed in German by the terms 'Entschiedungsfragen' and 'Ergänzungsfragen'. Alternatively we simply find references to questions of the first and second kind (e.g. in Kubiński (1966)).

[6] Many of the works of Hempel contain indications of the significance of 'why' questions for the analysis of scientific explanation.

[7] We understand assertion as a kind of epistemic modality. This means that in the analysis of assertion it is necessary to take account of who is making the assertion as well as of the asserted answer. Hence, assertion may be regarded as a function of two arguments $T\xi, \ldots$, where 'ξ' is the person making the assertion and '...' is the statement asserted. In what is to follow we shall be abstracting from the first of these two arguments. This may be achieved by, for instance, making the transition from 'ξ asserts that ...' to the unspecified 'it is asserted that ...'. We shall at the same time assume that there are epistemic reasons for

leaving out the first of the arguments, based on the fact that the person in question is qualified in respect of what is asserted and has enough evidence for what he is asserting, and so forth. For this abstracted form of assertion we shall be using the sign '⊦...'.

[8] It is not beyond the bounds of possibility that this situation might be expressed in terms of the devices of many-valued, or at least three-valued logic. Any such three-valued logic would, of course, have to allow for the possibility that the third value might be convertible to one of the first two.

[9] The latter alternative is important in the case of whether-questions which are not yes-or-no questions.

[10] The term 'use' is here meant in Quine's sense of 'to use' as opposed to 'to mention'.

[11] It should be pointed out that even Belnap avoids such a narrow conception.

[12] It may be of significance not only for assessing the meaningfulness of a question in research work, in the application of questionnaires, in a programmed teaching process etc., but also especially so for the automatization of the procedures involved in answering questions, as in the case of information storage and retrieval using a computer.

[13] The problem of ontic commitments was first outlined by Whitehead and Russell in the system of 'Principia Mathematica'. A more detailed treatment of the problem is to be found in Quine (1953a) and Quine (1960). On ontic commitments from the point of view of the use of terms see also Tondl (1966b).

[14] Russell gave this example in connection with the account of his conception of semantics and in connection with contradiction, but not in connection with the problem of questions, which he did not tackle.

[15] For more details on this kind of question see Giedymin (1964).

[16] Carnap's *C*-function serves this end (Carnap, 1950).

[17] The authors of some works on erotetic logic limit which-questions to 'which individual questions'. In the account to follow we shall be drawing attention to a further subgroup of which-questions, which we shall describe as 'which predicate questions'.

[18] If the class $(\hat{x})\ P_j x$ is empty the question is not meaningful, as we shall be showing in due course. It is of course quite possible to assume an infinite class, as has also been pointed out by Harrah (1963, p. 32) who speaks explicitly of 'an infinite, indefinite or relatively large set of possibilities'. However, in the case of an infinite class there will inevitably be difficulties as soon as the bounds of methodological finitism are overreached.

[19] E.g. Belnap (1963).

[20] The interpretation of the concept *L*-empty used here is based on Carnap's conception (Carnap 1956), including the introduction of meaning postulates conceived on the basis of a confirmed scientific theory.

[21] Hempel (1965, p. 334).

[22] Hempel (1965).

[23] This and any subsequent notations of why-questions assume the acceptance of the convention that inference rules undergo no change of validity in the assertion. This means, for example, that in

$$\begin{array}{l} \vdash (S_j \rightarrow S_i) \\ \vdash S_j \\ \hline \vdash S_i \end{array}$$

S_i is the legitimate conclusion which follows from the given premises. This convention is not of course acceptable for other epistemic modalities.

[24] This distinction is based on the emphasis laid on the semantic approach to the analysis of questions. This probably in turn means that the distinction is not able to embrace all

the refinements and shades of variety within the different types of questions in natural language and their day-to-day use in conversation. For instance, if we ask our friend: 'What made you decide ...?', the question may be interpreted as a which-question, e.g. 'what stimulus (out of a set of possible stimuli) made you decide ...?'. If we consider the sentence component 'you decide ...' to be unquestionably true because, say, a decision has been in fact made and can be verified, then the question may also be interpreted as a why-question: 'Why did you decide ...?'. And there is still the possibility of treating the whole question as a narrative question.

[25] In essence this is the empirical condition of the adequacy of scientific explanation. [One condition of this kind has been formulated by Hempel and Oppenheim (1948)]. This means that S_i, which is asserted in the explanandum, is a meaningful synthetic statement. If S_i is a logically truthful statement, then 'explaining' S_i has a different sense and is not of the nature of ontic decision.

[26] This characterization of the relations between the expressions which make up the explanans and those of the explanandum may appear very general. Conceptions of this relation have usually been formulated in a much simpler manner. In the well-known, classical work of Hempel and Oppenheim (1948) two fundamental conditions are given as the logical conditions of the adequacy of scientific explanation: (1) the explanandum must be the logical consequence of the explanans, (2) the explanans must involve general laws. According to these conditions scientific explanation is legitimate if there is a relation of logical consequence between the expressions which form the explanans and those which comprise the explanandum. Discussions around this presentation of scientific explanation, and the later works of Hempel [in particular Hempel (1962) and Hempel (1965)], have shown that this conception is too narrow and covers only the 'deductive-nomological model' of scientific explanation. However, even in the case of the 'statistical model' of scientific explanation it must be required that it be possible, on the basis of the validity of the expressions which make up the explanans, to draw conclusions as to the probability of what is asserted by the expressions in the explanandum.

[27] Hempel expresses this by saying that a scientific explanation must be 'potentially predictive' (Hempel 1965, p. 346).

[28] This description corresponds to Carnap's analysis of the procedure of explication (Carnap, 1950). This author distinguishes between the expression which is to be explicated (the explicandum) and that which gives the explication (the explicatum). The analogy with the pair explanandum–explanans, however, is merely exterior. It is first and foremost the logical structure of scientific explanation that is quite different. Furthermore it should be emphasized that an explanandum has the nature of an assertion, whereas an explicandum may be any kind of expression.

[29] See e.g. Pap (1962).

[30] The terms 'covering law' and 'model of explanation by covering law' were introduced by Dray (1957, 1963). The same terms are used in a similar sense by Hempel (1962, 1965).

[31] Hempel considers description and explanation to be the two main functions of the vocabulary of the language of science (Hempel, 1965; p. 139).

[32] From a familiar scene in the author's 'Le malade imaginaire'.

[33] This is the case of the example introduced in the section on problem-solving situations and questions where the question $(? \vdash)\ P_1 a_1$ is answered by $\vdash P_1' a_1$, the P_1 and P_1' being synonymous.

[34] Q_1 may be constituted by means of reduction sentences in Carnap's sense; see Carnap (1936).

[35] This is the way in which we might record the answer in the example given. There is of course another possible notation, expressing the relation of inclusion between the two

classes, i.e. $(\hat{x})\, P_1 x \subset (\hat{x})\, P_2 x$.
[36] Braithwaite (1960; p. 55 *et seq.*).
[37] Hempel and Oppenheim (1948) and numerous other studies, the most significant of which have been republished in the book *Aspects of Scientific Explanation* (Hempel 1965).
[38] Popper himself does not use this term.
[39] The term 'expression' is used here deliberately since there is always the possibility that a meaningful question $(?\vdash)(\vdash S_i)$ may contain a formula, theorem, etc., as the S_i. However, since the whole of our analysis relates primarily to the methodological problems of the empirical and experimental sciences, attention will be centred here mainly on situations where S_i is a meaningful synthetic statement.
[40] This case, where the explanandum is formulated as an assertion about the probability relations of certain properties which pertain to the mass phenomenon, should be kept distinct from cases where the formulation of the probability relations as statistical laws appears as a component in the composition of the explanans.
[41] This term was originally coined by Goodman (1947).
[42] The non-instrumentalist conception treats scientific laws as universal synthetic sentences. In contrast to this the instrumentalist conception denies that a scientific law says anything at all about the given universe. It understands a law merely as a 'calculational device', an adapted 'inferential rule' or 'inferential licence', i.e. rules which make it possible to deduce empirical data from other empirical data.
[43] There are of course yet more conditions. One requirement is, for example, that $(\forall x)(Px \rightarrow Qx)$ be what is known as a fundamental law and not merely a statement which can be derived from a fundamental law [see, e.g., Reichenbach (1947, p. 361), Hempel (1965, p. 267) among others]. Another is that the universe from which classes $(\hat{x})\, Px$ and $(\hat{x})Qx$ are taken be in no wise limited. This second requirement, which has evidently been adopted from physics, can be met very easily, though we would not be willing to give the respective general sentence the name of a 'law'. The sentence 'All objects that have the property denoted by P also have the property denoted Q' may relate to an unlimited universe, but we should not be willing to treat it as a law if there were only a single object with the property P.
[44] Carnap (1950; p. 294).
[45] This differentiation represents a certain modification of that to be found in Feigl (1953).
[46] The physical sciences tend to give preference to the designation 'dynamic laws' to underline the contrast with what are there called statistical laws.
[47] A more detailed analysis of the transitions between laws of the deterministic type and those of the statistical type has been made by Bohm (1957).
[48] This differentiation of laws according to the form of expression corresponds to the differences between classificatory, comparative and quantitative predicates.
[49] For this same distinction the terms 'dynamic laws' and 'structural laws' are sometimes also used. The former (from *dynamis*, i.e. power) is closely bound to certain ideas and principles of physics. The term 'structural laws' may also lead to a somewhat vague interpretation, since in view of the logical understanding of the concepts of structure and isomorphism even relations of time can be included among the other relevant relations.
[50] This question has been the topic of lengthy discussions. On the one side of the discussions stand those who support the simultaneous conception of cause and effect, and on the other those who favour the conception which links causality to a sequence of states.
[51] From the methodological point of view we must not overlook the problems associated with the 'simultaneous' measurement of two or more different characteristics of an object under investigation. 'Simultaneity' is of course no more than approximate, which is con-

nected with the statistical nature of measurement and possibly also with the need for the relativistic generalization of the concept of 'simultaneity'.
[52] Hempel and Oppenheim were not of course the first to expound the thesis of deductive inference in scientific explanation. Popper, in his *Logik der Forschung* (Popper, 1934) had pointed out earlier, in the course of the analysis of causal explanation, that giving a causal explanation of an event means deriving the statement describing that event from premises which contain one or more laws.
[53] These concepts may be considered as expressions for the facts or events of a dynamic system; particular facts or events can be described or expressed by a certain (finite) set of characteristics. In this respect, the concepts of 'cause' and 'effect' may be treated as *Substanzbegriffe*. However, it seems that the everyday way of understanding these expressions is better satisfied by treating them as *Relationsbegriffe*, i.e. two- or more-placed predicates roughly on the pattern: x is the cause of y under conditions z.
[54] We have in mind the requirement of the sufficiently certain or reliable possibility of prediction (retrodiction).
[55] Russell (1913).
[56] There are grounds for assuming that the degree of rationality may vary. There have even been attempts at categorizing various degrees, such as Weber's distinction of the *Zweckrational* and *Wertrational*, etc.
[57] See Nagel (1965).
[58] It is scarcely a coincidence that one of the first publications to expound some of the basic principles of cybernetics also related to the problems surrounding purpose, behaviour and teleology (Rosenblueth, Wiener and Bigelow, 1943).
[59] For this reason Nagel connects genetic explanation with the task of explanation in history. He also says that genetic explanation has effectively a narrative character since it gives an account of a sequence of unique events without explicitly stating or considering their causal connections (Nagel, 1965; pp. 264–268).
[60] Dray describes this mode of explanation as a 'continuous series of happenings' (Dray 1957; pp. 66–72).
[61] One reconstruction has, however, been attempted by Nagel (1965; p. 568).
[62] Schlick (1931).
[63] Ryle (1949; p. 121–123).
[64] Nor is there any general consensus as to how this standpoint should be designated. Reference is most commonly made here to the 'realistic' conception of scientific laws and theoretical predicates. Cf. e.g. Maxwell (1962) and Feyerabend (1958) among others.
[65] Nagel (1961).
[66] Stegmüller (1969; p. 97).
[67] Hempel (1965; p. 356).
[68] See in particular Scriven (1959, 1962).
[69] Scriven (1962; p. 196).
[70] Scriven (1959; p. 456).
[71] We use the term 'accidental generalization' for general statements to which we are not willing to concede the character of scientific laws. In the literature we find allusions in this sense to contingent or accidental statements. It is absolutely essential therefore to draw a strict distinction between accidental and empirical generalizations. Many major scientific discoveries originally had the character of empirical generalizations, i.e. empirically confirmed general statements, which, however, have still not been rooted in any definite theory. This is the case, for instance, of the original version of Kepler's laws, Boyle's laws and others. It should be stressed that, unlike accidental generalizations, empirical generalizations are capable of furnishing a relatively sound basis for explanation and

prediction.

[72] One of the first concise versions of this attempt was the work by Goodman, *The Problem of Counterfactual Conditionals* (Goodman, 1947). Here as in many other studies devoted to the subject, the problem is analyzed in much broader contexts. Counterfactual conditions are also examined in connection with dispositional predicates, the theory of confirmation, the conception of induction and inductive confirmation, among others.

[73] As mentioned above, the concept of 'lawlike statement' or 'lawlike sentence' was also introduced by Goodman (1947; p. 125). A lawlike statement is one having all the marks of a law except that its truthfulness is not determined. Scientific laws are therefore statements which are both lawlike and true. Naturally there are also statements which are true without being lawlike.

[74] A review and appreciation of these attempts is given by Stegmüller (1969; p. 283 *et seq.*).

[75] The matter here is not translation in the strict sense of the word, i.e. a transformation which also preserves the sense of the respective expressions. It is a matter of descriptions which have the same denotation in the given context. The procedure of replacing proper names by expressions having the character of descriptions is deemed fitting by some authors since it leads to the elimination of the semantic difficulties attached to proper names with zero denotation, such as 'Lucifer', 'Pegasus', etc.

[76] E.g. Hempel (1965; p. 267).

[77] The differentiation of basic, or original, laws and derived laws was first suggested by Reichenbach (1947; p. 361) who, however, conceived the differentiation more generally and spoke of original and derived nomological statements.

[78] Goodman (1947).

[79] The example is taken from Hempel (1965; p. 299). A very similar view has also been expressed by other writers. Popper (1934) early stressed that the formulations of scientific laws must contain 'purely universal predicates'.

[80] A more detailed account of the differences here is to be found in Švyrev's study (Švyrev, 1959).

[81] Pap (1962; p. 304).

[82] Reichenbach's conception appears in a more summary form in the 8th chapter of his *Elements of Symbolic Logic* (Reichenbach, 1947; p. 355 *et seq.*) and in the later work *Nomological Statements and Admissible Operations* (Reichenbach, 1954).

[83] Reichenbach (1947; p. 361).

[84] Reichenbach (1954; p. 7).

[85] Reichenbach (1947; p. 29). It should be stressed that for Reichenbach the concept 'connective operation' is the point of departure not only for the analysis of the concept of 'tautology' and 'tautological operations', e.g. 'tautological implication', which corresponds to Carnap's concept of *L*-implication, but also for the analysis of synthetic nomological statements, i.e. scientific laws.

[86] In *Elements of Symbolic Logic* (Reichenbach, 1947) he gives these requirements in a different order. The ordering we use proceeds from the simpler to the more demanding requirements.

[87] Reichenbach (1947, p. 368; and 1954, p. 29).

[88] He naturally makes no demand that every scientific law be expressible in such a language. He does consider, however, that his conception might be expanded to take in languages of a higher order (Reichenbach, 1954; p. 5).

[89] Reichenbach (1947; p. 264).

[90] In *Elements* (Reichenbach, 1947; p. 368) we find the expression 'demonstrable as true', or sometimes also the expression 'practically true'. In his later work we find only the ex-

pression 'verifiably true' (Reichenbach, 1954). This shows that Reichenbach had in mind the whole complex of operations capable of deciding the truth of a statement, also including some steps of a practical nature.

[91] Reichenbach (1954; p. 13).

[92] Reichenbach (1954; p. 18).

[93] See e.g. Lauter (1970; p. 140).

[94] E.g. Pap (1962; p. 301).

[95] Carnap (1950).

[96] Carnap's original conception of inductive confirmation, as we find it in *Logical Foundations of Probability* (Carnap, 1950), which attempted to quantify the degree of confirmation of a hypothesis h by the evidence available e on the basis of the meanings of h and e, was later subject to criticism, especially by Popper who referred back to his own conception of falsification (Popper, 1934). The main object of Popper's criticism was the identification of the concept of the 'degree of confirmation of sentence S_1 by sentence S_2' with the concept of 'conditional probability $p(S_1, S_2)$'. A summary of Popper's views on the degree of confirmation is to be found in Appendix IX to the last edition of his *Logik der Forschung* (Popper, 1969).

[97] This is not of course the only concept used to express the quality of a given general statement, i.e. in the main scientific law or hypothesis, with respect to given evidence. We also find, for instance, the concept 'likelihood', or 'relative acceptability'. Such concepts are usually based on the subjectivist conception of empirical knowledge.

[98] The main results and references to other works are to be found in the studies (Hintikka, 1968 and 1970; Pietarinen and Tuomela, 1968; and Pietarinen, 1970).

[99] Popper describes the origins of this idea in a note (Popper, 1969; p. 347) in which he draws attention to his priority.

[100] The term is coined by Pietarinen (1970).

[101] Bar-Hillel and Carnap (1953).

[102] Bar-Hillel and Carnap, in determining this measure of probability, used what they called the regular measure function, which they characterized with respect to the method of state description. The regular measure function m for the sentence S_i is the sum of the values m of all the state descriptions which make up the logical range of the statement S_i. Thus the content measure is defined as:

$$\text{cont}(S_i) = df\, 1 - m(S_i).$$

[103] Hintikka (1968; p. 316).

[104] Hintikka's conception of the concept 'transmitted information' relativizes the concept of information in the following sense. If we have a hypothesis S_h, then the information conveyed by the statement S_e with respect to what S_h refers to may be expressed in the form $\text{transinf}(S_e/S_h)$, where the sign 'transinf' means transmitted information. If S_e is a singular statement then it should be stressed that we are generally not interested merely in the fact to which the statement refers but also in other facts expressed in the general statement S_h, which are in some degree or other confirmed or refuted by the statement S_e. Herein lies the difference between $\text{inf}(S_e)$ and $\text{transinf}(S_e/S_h)$.

[105] In a paper given at the Vienna congress in 1968 (Pietarinen and Tuomela, 1969) reference is made to measures of explanatory power of a scientific theory. In a more extensive work of two years later (Pietarinen, 1970) the term 'systematic power' of a scientific hypothesis is used, the concept of 'systematization' including not only the procedures of explanation and prediction, but also certain empirical generalizations and those singular descriptive statements which are capable of bringing facts into a system, e.g. by locating them in a schema of classification.

[106] The uncertainty $U(S_e)$ may also be regarded as conditional uncertainty, taking it with respect to a statement S_t, which is an analytical statement (tautology). Then $U(S_e) = U(S_e/S_t)$.
[107] See Pietarinen and Tuomela (1969; p. 241), Pietarinen (1970, p. 125).
[108] This conception of scientific explanation is to be found in Campbell (1920) and Cohen and Nagel (1934).
[109] Popper (1934; esp. Chapter 12).
[110] Hempel and Oppenheim (1948).
[111] Hempel (1962).
[112] This requirement corresponds to that of what have been called nomological statements. However, we have already seen the difficulties of formulating exact and reliable criteria for nomological statements.
[113] A review of these attempts is to be found in Stegmüller (1969).
[114] E.g. Toulmin (1956).
[115] Hempel (1962).
[116] Hempel (1962).
[117] In this connection Hempel speaks of the epistemic relativity of statistical explanation (Hempel 1965; p. 402).
[118] One of the adherents of this conception is Reichenbach (1949; p. 375) who stressed that there exists but one legitimate conception of probability, namely the statistical frequency conception, which relates to classes and not to individual events.
[119] Carnap (1950) emphasizes that logical probability is a device which enables us to estimate an unknown value on the basis of evidence found, an instrument for decision-making, a 'guide of life'. Hence logical probability can be related to individual events, both present and future.
[120] A similar schema of inductive statistical explanation is given by Hempel, but only in the later works [esp. Hempel (1965, 1968)]. In his first work on statistical explanation, on the other hand, Hempel (1962) still uses schemata which hardly differ from those of inductive statistical explanation.
[121] Carnap (1950; p. 221).
[122] Humphreys gives a similar interpretation to the application of Carnap's requirement of total evidence in his criticism of Hempel's conception of inductive statistical explanation (Humphreys, 1968).
[123] Carnap (1950; Ch. VI).
[124] Condition (2) is formulated differently from that in Carnap (1950; p. 348). On principle Carnap takes as arguments of his *c*-functions sentences, not predicates. Thus his conditions are formulated for the relevance or irrelevance of sentences while in our conception we are considering the relevance of predicates. Carnap's formulation of condition (d) runs: i is irrelevant to h on the basis of the evidence $e =df$ either (1) $c(h, e.i) = c(h, e)$ or (2) e, i is *L*-false.
[125] Supporters of the classical deterministic conception in physics assume that some such determinization of a statistical hypothesis is always at least theoretically possible. The elements of the class $\{P_1, P_2, \ldots, P_n\}$ which make the determinization of the statistical hypothesis possible may be described as 'hidden parameters'. One necessary objection to this conception is that in science there is never any *a priori* guarantee that the class $\{P_1, P_2, \ldots, P_n\}$ actually contains any hidden parameters. The impossibility of the existence of any such guarantee also follows from the fact that, from the viewpoint of methodological finitism, the class $\{P_1, P_2, \ldots, P_n\}$ must always be assumed to be finite, although the possibility of its being extended (finitely) is never excluded.
[126] Hempel (1962).

[127] Hempel (1965; p. 400; 1968, p. 120).
[128] Wójćicki (1966). Hempel later accepted the criticism in full, see Hempel (1968; p. 126).
[129] Humphreys (1968).
[130] Hempel (1968; p. 124 and 127).
[131] Stegmüller uses the term 'modal qualifier' in the same sense (Stegmüller, 1969; p. 650).
[132] Stegmüller (1969; p. 650).
[133] Stegmüller (1969; p. 702).
[134] This concept was introduced in the work Perez and Tondl (1965).
[135] Ledley and Lusted (1959).
[136] Hempel has already pointed out these important differences (Hempel, 1965; pp. 375 and 460). A more detailed account of the difference between causal concepts and the concept of 'symptoms' has been given by Stegmüller, who points out that explanation in the proper sense of the word can only operate with what he describes as *Ursachen oder Realgründe*, and not with what might be described as *Vernunftgründe* (Stegmüller, 1969; pp. 190–192). Wherever we operate with symptoms in any such procedure, they have the character of *Vernunftgründe*.
[137] See for example Wald (1950) and Blackwell and Girshick (1954).
[138] Wald (1950).
[139] On this conception of statistical decision see, for example, Blackwell and Girshick (1954).
[140] Since the endeavour of the whole of this book is to further the informational view of scientific procedures, i.e. the conception which regards scientific procedures as operations with data, we also apply the terminology currently used in the analysis of communication processes.
[141] See for example Wald (1950; Chapter 1 and 4).
[142] For a more detailed account of the problems of reduction in scientific procedures see Perez and Tondl (1965).
[143] On the ambiguity of the expression 'is' from the point of view of chronological logic, see Rescher (1969; p. 158).
[144] The term 'prophesy' is sometimes used in the literature to mean those prognostic statements that relate to systems which man is not able to influence in any way. In contrast, prognostic statements relating to systems where man can make his activity felt, are generally described as 'technical predictions'. [This differentiation can be found, for example in Popper (1957)]. However I do not consider these connotations to be entirely suited since certain extremely accurate and dependable predictions, such as in astronomy or meteorology, would have to be treated as 'prophesies'. This is not of course to deny the usefulness of distinguishing technical predictions from predictions relating to systems whose evolution we are not able to influence. A detailed semantic analysis and classification of prognostic statements exceeds the scope of the present work.
[145] Hempel and Oppenheim (1948; p. 332).
[146] Hempel (1965; p. 367).
[147] Hempel introduced the concept of 'systematization' in his work *The Theoretician's Dilemma*, which appeared in Hempel (1965; p. 174).
[148] Hempel of course did not understand his concept of 'systematization' as broadly as this. It can, however, be shown fairly easily that in cases of derived measurement (e.g. determining the temperature of a gas if we know the pressure and volume and the appropriate constant) based on a nomological statement (in this case, Boyle's law) we proceed along the lines of the schema which corresponds to Hempel's deductive nomological model of scientific explanation. It might also be said that derived measuring described in this way is of the nature of what we have described as codiction.

[149] This objection is similar to that raised against the thesis of structural similarity by Scheffler (1964) who distinguished between causal explanation and the explanation of opinions, beliefs, attitudes, etc. The theory of structural similarity can in no sense relate to the latter group of explanations, unless of course we deny the possibility of their being 'adequately explained'.
[150] Stegmüller (1969; p. 171).
[151] Both types of why-questions given here are of the character of 'epistemic why-questions', since in neither case is the basis of the question a true statement.
[152] Stegmüller says that in the first case what is required is the statement of causes, or what he calls *Seinsgründe* or *Realgründe*, while in the second case only *Vernunftsgründe* are called for.
[153] A review of these prognostic methods is to be found in Jantsch's publication (Jantsch, 1967). In addition to the Delphi method there are a number of other methods based on the judgment of competent experts, e.g. 'brainstorming', 'scenario', and others.
[154] Hempel (1965; p. 372) describes explanations of this kind as 'self-evidencing explanations'. He points out that although self-evidencing explanations have an exterior form corresponding to the deductive nomological model, they are in effect incomplete explanations, the explanans not fulfilling what is normally required of it.
[155] Wiener (1947; Chapter 1).
[156] A noteworthy application of this approach in the analysis of time in economics is to be found in the studies of Hrubý (1965 and 1970).
[157] The concept of 'predictive basis' corresponds to Hempel's concept of 'potential explanans' (Hempel, 1965; p. 277 and p. 293). On some of the difficulties encountered in attempts to define this concept with greater accuracy see Stegmüller (1969; p. 708 *et seq.*).
[158] The given schema, which is based on the difference of two multiples, is nothing new to the literature on inductive logics. A similar schema, based however on the gain or loss associated with the nomological statement S_h, has been introduced by Pietarinen (1970; p. 140).

BIBLIOGRAPHY

Åqvist, Lennart, *A New Approach to the Logical Theory of Interrogatives*, Uppsala 1965.

Bar-Hillel, Y. and Carnap, R., 'Semantic Information', in *Communication Theory* (ed. by W. Jackson), London 1953.

Belnap, N. D., 'An Analysis of Questions', Preliminary Report, Technical Memorandum Series, 1963.

Blackwell, D. and Girshick, M. A., *Theory of Games and Statistical Decisions*, New York 1954.

Bohm, David, *Causality and Chance in Modern Physics*, London 1957.

Braithwaite, R. B., *Scientific Explanation: a Study of the Function of Theory, Probability and Law in Science*, Harper Torchbooks, 1960, New York, Evanston, and London.

Bridgman, P. W., 'The Nature of Some of Our Physical Concepts', *The British Journal for the Philosophy of Science* (1951), Vol. I, pp. 257–272, Vol. II, pp. 25–44.

Brillouin, Léon, *Science and Information Theory*, New York 1962.

Buchdahl, Gerd, 'Semantic Sources of the Concept of Law', *Synthese* **17**, No. 1 (1967).

Campbell, N. R., *Physics: the Elements*, Cambridge Univ. Press, London, 1920.

Campbell, N. R., *An Account of the Principles of Measurements and Calculations*, London 1928.

Carnap, Rudolf, *Der Logische Aufbau der Welt*, Berlin 1928. (Zweite Aufl., Hamburg 1961.)

Carnap, Rudolf, *Logische Syntax der Sprache*, Wien 1934.

Carnap, Rudolf, 'Testability and Meaning' (Reprinted from *Philosophy of Science*, Vol. III, 1936, and Vol. IV, 1937; with corrigenda and additional bibliography by Prof. Carnap). Graduate Philosophy Club, Yale University, New Haven, Conn. 1950.

Carnap, Rudolf, *Logical Foundations of Probability*, Chicago 1950.

Carnap, Rudolf, *Meaning and Necessity*, 2nd ed., Chicago 1956(a).

Carnap, Rudolf, 'The Methodological Character of Theoretical Concepts', *Minnesota Studies in the Philosophy of Science* **1**, Minneapolis 1956(b).

Carnap, Rudolf, *Philosophical Foundations of Physics*, New York 1966.

Cohen, M. R. and Nagel, E., *An Introduction to Logic and Scientific Method*, New York 1934.

Dray, W., *Laws and Explanation in History*, Oxford 1957.

Dray, W., 'The Historical Explanation of Actions Reconsidered', in *Philosophy and History* (ed. by S. Hook), New York 1963.

Feigl, Herbert, 'Notes on Causality', in *Readings in the Philosophy of Science* (ed. by H. Feigl and M. Bradbeck), New York 1953.

Feyerabend, P. K., 'An Attempt at a Realistic Interpretation of Experience', *Proceedings of the Aristotelian Society* **58** (1958) pp. 144–170.

Feyerabend, P. K., 'Explanation, Reduction and Empiricism', in *Minnesota Studies in the Philosophy of Science* (ed. by G. Feigl and G. Maxwell), Vol. III. Minneapolis 1962.

Frank, Philipp, *Philosophy of Science*, Englewood Cliffs, N.J., 1957.

Gabor, D., 'Theory of Communication', *Journal of the Inst. Elec. Engrs.* **93** (1946) Part. III, London.

Giedymin, Jerzy, *Problemy, Założenia, Rozstrzygniecia*, Studia nad logicznymi podstawami

nauk spolecznych, Poznań 1964.
Goodman, Nelson, 'The Problem of Counterfactual Conditionals', *The Journal of Philosophy* **44** (1947), pp. 113–128.
Goodman, Nelson, *The Structure of Appearence*, Cambridge, Mass. 1951.
Goodman, Nelson, *Fact, Fiction and Forecast*, Cambridge, Mass. 1955.
Harrah, David, 'A Logic of Questions and Answers', *Philosophy of Science* **28** (1961) pp. 40–46.
Harrah, David, *Communication: A Logical Model*, Cambridge, Mass. 1963.
Hempel, C. G. and Oppenheim, P., 'The Logic of Explanation', *Philosophy of Science* **15** (1948).
Hempel, C. G., 'Fundamentals of Concept Formation in Empirical Sciences', *International Encyclopedia of Unified Science* **II**, Chicago 1952.
Hempel, C. G., 'Deductive-Nomological vs. Statistical Explanation', *Minnesota Studies in the Philosophy of Science* **III**, Minneapolis 1962.
Hempel, C. G., 'Implications of Carnap's Work for the Philosophy of Science', in *The Philosophy of Rudolf Carnap* (ed. by P. Schilpp), 1963.
Hempel, C. G., *Aspects of Scientific Explanation*, New York 1965.
Hempel, C. G., 'Maximal Specificity and Lawlikeness in Probabilistic Explanation', *Phil. of Science* **32**, (1968).
Hintikka, Jaako, 'The Varieties of Information and Scientific Explanation', in *Logic, Methodology and Philosophy of Science III* (ed. by B. V. Rootselaar and J. F. Staal), Proceedings of the Third Inter. Congress for Logic, Methodology and Phil. of Science, Amsterdam 1968.
Hintikka, Jaako, 'On Semantic Information', in *Information and Inference* (ed. by J. Hintikka and P. Suppes), Dordrecht 1970.
Hrubý, Pavel, *Teorie ekonomického času, Politická ekonomie*, Praha 1969, č. 7.
Hrubý, Pavel, *Dlouhodobé trendy ekonomického času, Politická ekonomie*, Praha 1970, č. 5.
Humphreys, W. C., 'Statistical Ambiguity and Maximal Specificity', *Philosophy of Science* **35**, No. 2, June 1968.
Jantsch, Erich, *Technological Forecasting in Perspective*, OECD, Paris 1967.
Juhos, von Bela, 'Über die Definierbarkeit und empirische Anwendung von Dispositionsbegriffen', *Kant-Studien*, Bd. 51, H.3, Köln 1959–1960.
Kedrov, B. and Spirkin, A., 'Nauka', in *Filosofskaja enciklopedija*, t. 3, Moskva 1964.
Kopnin, P. V., 'O napravlenijach v razrabotke logiki nauki', in *Logika i metodologija nauki*, Moskva 1967.
Körner, Stephan, *Conceptual Thinking; A Logical Inquiry*, New York 1959.
Körner, Stephan, *Experience and Theory*, London 1966.
Kubiński, T., 'An Essay in Logic of Questions', *Atti del XII. Congresso Inter. di Filosofia* (Venezia 1958), Vol. 5, pp. 315–322.
Kubiński, T., 'Przegląd niektórych zagednień logiki pytań', *Studia logica*, t. XVIII, Warszawa-Poznań 1966.
Lauter, H. A., 'An Examination of Reichenbach on Laws', *Philosophy of Science* **37**, (1970).
Ledley, R. S., Lusted, L. B., 'Reasoning Foundations of Medical Diagnosis', *Science* **130** (1959) pp. 9–21.
Lednikov, E. E., 'Ponjatie 'konstrukt' v logike nauki', *Logika i metodologija nauki*, Moskva 1967.
Maxwell, Grover, 'The Ontological Status of Theoretical Entities', in *Minnesota Studies in the Philosophy of Science* (ed. by H. Feigl and G. Maxwell), Vol. III, Minneapolis 1962.

Menger, Karl, 'Mensurations and other Mathematical Connections of Observable Material', in *Definitions and Theories* (ed. by C. W. Churchman and Ph. Ratoosh), New York 1959.
Nagel, Ernest, 'Teleological Explanation and Teleological Systems', in *Readings in the Philosophy of Science* (ed. by H. Feigl and M. Brodbeck), New York 1953.
Nagel, Ernest, *The Structure of Science: Problems in the Logic of Scientific Explanation*, New York 1961.
Narskij, I. S., 'Dispozicionnyje predikaty i problema tak nazyvajemych vtoryčnych kačestv', in *Logika i metodologija nauki*, Moskva 1967.
Nikiforov, A. L., 'O formalnych priznakach utverždenij vyražajuščich naučnyje zakony', *Voprosy filosofii* No. 11 (1968).
Pap, Arthur, *Elements of Analytic Philosophy*, New York 1949.
Pap, Arthur, *An Introduction to the Philosophy of Science*, New York 1962.
Perez, Albert and Tondl, Ladislav, 'Modely některých vědeckých procedur z hlediska logiky a teorie informace', in *Problémy kybernetiky*, (ed. by J. Kožešník), Nakl. ČSAV, Praha 1965.
Perez, Albert and Tondl, Ladislav, 'On the Role of Information Theory in Certain Scientific Procedures', in *Information and Prediction in Science* (ed. by S. Dockx and P. Bernays), New York 1965.
Pietarinen, Juhani and Tuomela, Raimo, 'On Measures of the Explanatory Power of Scientific Theories', in *Akten des XIV. Internationalen Kongresses für Philosophie*, Wien 2–9. September 1968, Wien 1968.
Pietarinen, Juhani, 'Quantitative Tools for Evaluating Scientific Systematizations', in *Information and Inference* (ed. by J. Hintikka and P. Suppes), Dordrecht 1970.
Polikarov, A., 'O metodologii rozwiązywania problemów naukowych', *Studia filozoficzne*, No. 2, 1967.
Popovič, M. V., 'O konstruktivnom podchodě k analizu jazyka nauki', in *Logika i metodologija nauki*, Moskva 1967.
Popper, Karl R., *Logik der Forschung*, 1. Aufl., Wien 1934.
Popper, Karl R., *The Poverty of Historicism*, New York 1957.
Popper, Karl R., *The Logic of Scientific Discovery*, London 1959.
Popper, Karl R., *Logik der Forschung*, 3. Aufl., Tübingen 1969.
Putnam, Hilary, 'What Theories Are Not', in *Logic, Methodology and Philosophy of Science* (ed. by E. Nagel, P. Suppes, end A. Tarski), Proceedings of the 1960 Int. Congress, Stanford 1962.
Quine, Willard van Orman, 'Notes on Existence and Necessity', *Journal of Philosophy* **XI** (1943) pp. 113–123. Repr. in *Semantics and the Philosophy of Language* (ed. by L. Linsky), Urbana 1952.
Quine, Willard van Orman, 'On What There Is', in *From a Logical Point of View*, Cambridge, Mass. 1953 (a).
Quine, Willard van Orman, 'Logic and the Reification of Universals', in *From a Logical Point of View*, Cambridge, Mass. 1953 (b).
Quine, Willard van Orman, *Word and Object*, Cambridge, Mass. 1960.
Reichenbach, Hans, *Experience and Prediction*, Chicago 1938.
Reichenbach, Hans, *Elements of Symbolic Logic*, New York 1947.
Reichenbach, Hans, *The Theory of Probability*, Berkeley and Los Angeles 1949.
Reichenbach, Hans, *Nomological Statements and Admissible Operations*, Amsterdam 1954.
Reichenbach, Hans, *The Direction of Time*, Berkeley and Los Angeles 1956.
Reichenbach, Hans *The Philosophy of Space and Time*, New York 1958.

Rosenblueth, Arturo, Wiener, Norbert, and Bigelow, Julian, 'Behavior, Purpose and Teleology', *Philosophy of Science* **10** (1943).
Russell, Bertrand, 'On Denoting', *Mind* **XIV** (1905), str. 479–493.
Russell, Bertrand, 'On the Notion of Cause', *Proceedings of the Aristotelian Society* **XIII**, pp. 1–13. Repr. in *Mysticism and Logic*, New York 1918.
Russell, Bertrand, *Introduction to Mathematical Philosophy*, London 1919.
Russell, Bertrand, *Inquiry into Meaning and Truth*, London 1940.
Russell, Bertrand, *Human Knowledge: Its Scope and Limits*, London-New York 1948.
Ryle, Gilbert, *The Concept of Mind*, London 1949.
Ryle, Gilbert, 'The Theory of Meaning', in *British Philosophy in the Mid-Century*, (ed. by C. A. Mace), London 1957.
Scheffler, Israel, *The Anatomy of Inquiry*, New York and London 1964.
Schlick, Moritz, 'Die Kausalität in der gegenwärtigen Physik', *Die Naturwissenschaften*, Bd. 19 (1931), pp. 145–162.
Scriven, Michael, 'Truisms as Grounds for Historical Explanation', in *Theories of History* (ed. by P. Gardiner), New York 1959.
Scriven, Michael, 'Explanation, Prediction, and Laws', in *Minnesota Studies in the Philosophy of Science* (ed. by H. Feigl and G. Maxwell), Vol. III, Minneapolis 1962.
Simmons, Robert F., 'Answering English Questions by Computer', in *The Growth of Knowledge* (ed. by Mandred Kochen), New York 1967.
Smart, J. J. C., 'Conflicting Views about Explanation', in *Boston Studies in the Philosophy of Science*, Dordrecht and New York 1965.
Stegmüller, Wolfgang, *Wissenschaftliche Erklärung und Begründung*, Berlin-Heidelberg-New York 1969.
Švyrev, V. S., 'K voprosu o kauzal'noj implikacii', in *Logičeskije issledovanija*, Moskva 1959.
Tarski, Alfred, *Introduction to Logic* (Ninth print), New York 1961.
Tavanec, P. V. and Švyrev, V. S. 'Logika naučnogo poznanija', in *Problemy logiki naučnogo poznanija*, Moskva 1967.
Tondl, Ladislav, 'Antinomy of 'Liar' and Antinomy of Synonymous Names', *Kybernetika*, No. 1, 1966 (a).
Tondl, Ladislav, *Problémy sémantiky*, Academia, Praha 1966 (b).
Tondl, Ladislav, *Člověk a věda*, Academia, Praha 1969.
Toulmin, S., *The Uses of Argument*, Cambridge 1958.
Uemov, A. I., *Věšči, svoistva, otnošenija*, Moskva 1963.
Vigier, J. P., 'Determinism and Indeterminism in a New 'Level' Theory of Matter', in *Logic, Methodology and Philosophy of Science* (ed. by E. Nagel and P. Suppes), Stanford 1962, pp. 262–264.
Vigier, J. P., 'Teorija urovnej i dialektika prirody', *Voprosy filosofii*, No. 10, 1962.
Wald, Abraham, *Statistical Decision Functions*, New York 1950.
Watanabe, S., 'Une explication mathématique du classement d'objects', in *Information and Prediction in Science*, (ed. by S. Dockx and P. Bernays), New York 1965.
Wiener, Norbert, *The Human Use of Human Beings, Cybernetics and Society*, London 1954.
Wittgenstein, Ludwig, *Tractatus logico-philosophicus*, New York and London 1922.
Wójćicki, Ryszard, 'Filozofia nauki w 'Minnesota Studies'', *Studia filozoficzne* (1966), pp. 143–154.
Wright, Georg H. von, *An Essay in Modal Logic*, Amsterdam 1951.
Žarikov, E. S., 'Problema predskazyvanija v nauke', in *Logika i metodologija nauki*, Moskva 1967.
Zinovjev, A. A., 'O vozmožnostjach logičeskogo analiza nauki', in *Logika i metodologija nauki*, Moskva 1967.

SYNTHESE LIBRARY

Monographs on Epistemology, Logic, Methodology,
Philosophy of Science, Sociology of Science and of Knowledge, and on the
Mathematical Methods of Social and Behavioral Sciences

MARTIN STRAUSS, *Modern Physics and Its Philosophy*. 1972, X+297 pp. Dfl. 80,—

SÖREN STENLUND, *Combinators, λ-Terms and Proof Theory*. 1972, 184 pp. Dfl. 49,—

DONALD DAVIDSON and GILBERT HARMAN (eds.), *Semantics of Natural Language*. 1972, X+769 pp. (Cloth) Dfl. 95,— (Paper) Dfl. 45,—

‡STEPHEN TOULMIN and HARRY WOOLF (eds.), *Norwood Russell Hanson: What I Do Not Believe, and Other Essays*. 1971, XII+390 pp. Dfl. 90,—

‡ROGER C. BUCK and ROBERT S. COHEN (eds.), *PSA 1970. In Memory of Rudolf Carnap. Boston Studies in the Philosophy of Science*. Volume VIII (ed. by Robert S. Cohen and Marx W. Wartofsky). 1971, LXVI+615 pp. (Cloth) Dfl. 120,— (Paper) Dfl. 60,—

‡YEHOSHUA BAR-HILLEL (ed.), *Pragmatics of Natural Languages*. 1971, VII+231 pp. Dfl. 50,—

‡ROBERT S. COHEN and MARX W. WARTOFSKY (eds.), *Boston Studies in the Philosophy of Science*. Vol. VII: Milič Čapek: *Bergson and Modern Physics*. 1971, XV+414 pp. Dfl. 70,—

‡CARL R. KORDIG, *The Justification of Scientific Change*. 1971, XIV+119 pp. Dfl. 33,—

‡JOSEPH D. SNEED, *The Logical Structure of Mathematical Physics*. 1971, XV+311 pp. Dfl. 70,—

‡JEAN-LOUIS KRIVINE, *Introduction to Axiomatic Set Theory*. 1971, VII+98 pp. Dfl. 28,—

‡RISTO HILPINEN (ed.), *Deontic Logic: Introductory and Systematic Readings*. 1971, VII+182 pp. Dfl. 45,—

‡EVERT W. BETH, *Aspects of Modern Logic*. 1970, XI+176 pp. Dfl. 42,—

‡PAUL WEINGARTNER and GERHARD ZECHA (eds.), *Induction, Physics, and Ethics. Proceedings and Discussions of the 1968 Salzburg Colloquium in the Philosophy of Science*. 1970, X+382 pp. Dfl. 65,—

‡ROLF A. EBERLE, *Nominalistic Systems*. 1970, IX+217 pp. Dfl. 42,—

‡JAAKKO HINTIKKA and PATRICK SUPPES, *Information and Inference*. 1970, X+336 pp. Dfl. 60,—

‡KAREL LAMBERT, *Philosophical Problems in Logic. Some Recent Developments*. 1970, VII+176 pp. Dfl. 38,—

‡P. V. TAVANEC (ed.), *Problems of the Logic of Scientific Knowledge*. 1969, XII+429 pp. Dfl. 95,—

‡ROBERT S. COHEN and RAYMOND J. SEEGER (eds.), *Boston Studies in the Philosophy of Science*. Volume VI: *Ernst Mach: Physicist and Philosopher*. 1970, VIII+295 pp. Dfl. 38,—

‡MARSHALL SWAIN (ed.), *Induction, Acceptance, and Rational Belief*. 1970, VII+232 pp. Dfl. 40,—

‡NICHOLAS RESCHER *et al.* (eds.), *Essays in Honor of Carl G. Hempel. A Tribute on the Occasion of his Sixty-Fifth Birthday*. 1969, VII+272 pp. Dfl. 50,—

‡PATRICK SUPPES, *Studies in the Methodology and Foundations of Science. Selected Papers from 1951 to 1969*. 1969, XII+473 pp. Dfl. 72,—

‡JAAKKO HINTIKKA, *Models for Modalities. Selected Essays*. 1969, IX+220 pp. Dfl. 34,—

‡D. DAVIDSON and J. HINTIKKA (eds.), *Words and Objections: Essays on the Work of W. V. Quine*. 1969, VIII+366 pp. Dfl. 48,—

‡J. W. DAVIS, D. J. HOCKNEY and W. K. WILSON (eds.), *Philosophical Logic*. 1969, VIII+277 pp. Dfl. 45,—

‡ROBERT S. COHEN and MARX W. WARTOFSKY (eds.), *Boston Studies in the Philosophy of Science*. Volume V: *Proceedings of the Boston Colloquium for the Philosophy of Science 1966/1968*. 1969, VIII+482 pp. Dfl. 60,—

‡ROBERT S. COHEN and MARX W. WARTOFSKY (eds.), *Boston Studies in the Philosophy of Science*. Volume IV: *Proceedings of the Boston Colloquium for the Philosophy of Science 1966/1968*. 1969, VIII+537 pp. Dfl. 72,—

‡NICHOLAS RESCHER, *Topics in Philosophical Logic*. 1968, XIV+347 pp. Dfl. 70,—

‡GÜNTHER PATZIG, *Aristotle's Theory of the Syllogism. A Logical-Philological Study of Book A of the Prior Analytics*. 1968, XVII+215 pp. Dfl. 48,—

‡C. D. BROAD, *Induction, Probability, and Causation. Selected Papers*. 1968, XI+296 pp. Dfl. 54,—

‡ROBERT S. COHEN and MARX W. WARTOFSKY (eds.), *Boston Studies in the Philosophy of Science*. Volume III: *Proceedings of the Boston Colloquium for the Philosophy of Science 1964/1966*. 1967, XLIX+489 pp. Dfl. 70,—

‡GUIDO KÜNG, *Ontology and the Logistic Analysis of Language. An Enquiry into the Contemporary Views on Universals*. 1967, XI+210 pp. Dfl. 41,—

*EVERT W. BETH and JEAN PIAGET, *Mathematical Epistemology and Psychology*. 1966, XXII+326 pp. Dfl. 63,—

*EVERT W. BETH, *Mathematical Thought. An Introduction to the Philosophy of Mathematics*. 1965, XII+208 pp. Dfl. 37,—

‡PAUL LORENZEN, *Formal Logic*. 1965, VIII+123 pp. Dfl. 26,—

‡GEORGES GURVITCH, *The Spectrum of Social Time*. 1964, XXVI+152 pp. Dfl. 25,—

‡A. A. ZINOV'EV, *Philosophical Problems of Many-Valued Logic*. 1963, XIV+155 pp. Dfl. 32,—

‡MARX W. WARTOFSKY (ed.), *Boston Studies in the Philosophy of Science*. Volume I: *Proceedings of the Boston Colloquium for the Philosophy of Science, 1961–1962*. 1963, VIII+212 pp. Dfl. 26,50

‡B. H. KAZEMIER and D. VUYSJE (eds.), *Logic and Language. Studies Dedicated to Professor Rudolf Carnap on the Occasion of his Seventieth Birthday*. 1962, VI+256 pp. Dfl. 35,—

*EVERT W. BETH, *Formal Methods. An Introduction to Symbolic Logic and to the Study of Effective Operations in Arithmetic and Logic*. 1962, XIV+170 pp. Dfl. 35,—

*HANS FREUDENTHAL (ed.), *The Concept and the Role of the Model in Mathematics and Natural and Social Sciences. Proceedings of a Colloquium held at Utrecht, The Netherlands, January 1960*. 1961, VI+194 pp. Dfl. 34,—

‡P. L. R. GUIRAUD, *Problèmes et méthodes de la statistique linguistique*. 1960, VI+146 pp. Dfl. 28,—

*J. M. BOCHEŃSKI, *A Precis of Mathematical Logic*. 1959, X+100 pp. Dfl. 23,—

SYNTHESE HISTORICAL LIBRARY

Texts and Studies
in the History of Logic and Philosophy

‡LEWIS WHITE BECK (ed.), *Proceedings of the Third International Kant Congress, held at the University of Rochester, March 30–April 4, 1970*. 1972, XI + 718 pp. Dfl. 160,—

‡KARL WOLF and PAUL WEINGARTNER (eds.), *Ernst Mally: Logische Schriften*. 1971, X + 340 pp. Dfl. 80,—

‡LEROY E. LOEMKER (ed.), *Gottfried Wilhelm Leibniz; Philosophical Papers and Letters*. A Selection Translated and Edited, with an Introduction. 1969, XII + 736 pp. Dfl. 125,—

‡M. T. BEONIO-BROCCHIERI FUMAGALLI, *The Logic of Abelard*. Translated from the Italian. 1969, IX + 101 pp. Dfl. 27,—

Sole Distributors in the U.S.A. and Canada:
*GORDON & BREACH, INC., 440 Park Avenue South, New York, N.Y. 10016
‡HUMANITIES PRESS, INC., 303 Park Avenue South, New York, N.Y. 10010